SpringerBriefs in Environmental Science

Photo by John Akerman

SpringerBriefs in Environmental Science present concise summaries of cutting-edge research and practical applications across a wide spectrum of environmental fields, with fast turnaround time to publication. Featuring compact volumes of 50 to 125 pages, the series covers a range of content from professional to academic. Monographs of new material are considered for the SpringerBriefs in Environmental Science series.

Typical topics might include: a timely report of state-of-the-art analytical techniques, a bridge between new research results, as published in journal articles and a contextual literature review, a snapshot of a hot or emerging topic, an in-depth case study or technical example, a presentation of core concepts that students must understand in order to make independent contributions, best practices or protocols to be followed, a series of short case studies/debates highlighting a specific angle.

SpringerBriefs in Environmental Science allow authors to present their ideas and readers to absorb them with minimal time investment. Both solicited and unsolicited manuscripts are considered for publication.

More information about this series at http://www.springer.com/series/8868

Luciana S. Esteves

Managed Realignment: A Viable Long-Term Coastal Management Strategy?

Springer

Luciana S. Esteves
Bournemouth University
Faculty of Science and Technology
Poole
UK

ISSN 2191-5547   ISSN 2191-5555 (electronics)
ISBN 978-94-017-9028-4   ISBN 978-94-017-9029-1 (eBook)
DOI 10.1007/978-94-017-9029-1
Springer Dordrecht Heidelberg New York London

Library of Congress Control Number: 2014940853

Printed on acid-free paper

Springer is part of Springer Science+Business Media (www.springer.com)

Preface

While I am writing this preface, long spells of extreme weather are affecting people worldwide. Stormy and wet weather have caused floods that are disrupting lives across England and Wales for more than three months. The Met Office has announced that January 2014 was the wettest in some parts of the UK since records began to be collected more than 100 years ago. Heat waves and record high temperatures mark the summer in southern Australia and southern Brazil. The 'big freeze' affects the USA where some locations have had the coldest temperatures ever recorded. Due to population growth and climate change, the challenges brought by extreme weather will be faced by more and more people each year.

This is particularly concerning in coastal areas, where extreme storms can coincide with high tides and exacerbate flooding and erosion risk to people, economies and natural habitats. Examples of such extreme conditions are widespread, such as seen in the USA during hurricanes Katrina (New Orleans in 2005) and Sandy (New York in 2012), in the Philippines during Tropical Storm Trami (Manila in 2013) and typhoon Haiyan (Tacloban in 2013) and the recent floods in the UK (affecting East Anglia, Southwest, Midlands and Wales), just to name a few. No single event can be attributed to climate change. Independent from whether climate change is human-induced or not, the effects of climate change are upon us. Adaptation is inevitable and we all (individually and collectively) need to learn new ways of living to become more resilient to the consequences of extreme weather events.

This book is not about extreme weather or climate change. It is about how they are provoking a change in the way flooding and erosion risks are managed in coastal areas and the alternatives that exist to counteract their negative socio-economic and environmental effects. More specifically, this book describes and discusses a relatively new alternative called managed realignment. Managed realignment is a soft engineering approach that aims to provide a more sustainable way to manage coastal erosion and flood risk by enhancing the natural adaptive capacity of intertidal habitats. In other words, managed realignment creates space for coastlines to evolve more naturally, adjusting dynamically to changing environmental conditions, including rising sea levels. Therefore, managed realignment represents a shift from the traditional hard engineering approach to coastal protection.

Although managed realignment is becoming increasingly popular worldwide, its implementation is faced by great challenges, including public acceptance, limited knowledge, funding constraints and uncertainties related to natural coastal evolution. The first managed realignment projects were implemented in the 1980s (in France, Germany and the Netherlands). In the early 2000s, managed realignment became an important climate change adaptation mechanism in national strategies (e.g. in the UK) aiming to deliver economically and environmentally sustainable coastal management. Since then, there has been a great increase in the number of publications addressing the subject. However, the great majority are 'white papers' produced by government agencies and consultants involved in the design or delivery of managed realignment projects. Only few independent studies have been published and even fewer present reliable documentation and analysis of how realigned sites are actually evolving.

This is the first book focusing on managed realignment. It is written with the objective to provide an independent overview about managed realignment, how it has been implemented in different countries, its current achievements and limitations, and its potential to deliver sustainable long-term coastal management. Investments in managed realignment are increasing. In the UK, there are plans to realign 10 % of the English and Welsh coastline by 2030. In France, the *National Strategy for Integrated Shoreline Management* explicitly includes planning for retreat from high risk areas as a priority action. It is time that managed realignment is clearly explained and independently assessed.

The content of the book provides a balance between academic research and practical experience. I have used my academic judgement based on the available literature to clarify the basic concepts and definitions and summarise the state-of-the-art knowledge about managed realignment. In five chapters, external contributions provide an account of practical experience in planning, designing and implementing managed realignment in the Netherlands, the USA and the UK. The book describes general concepts and discusses approaches used internationally. However, the content does have a European and British bias, as most of the literature available is produced in Europe, and the UK is the country with the largest number of known projects.

The first three chapters provide an overview of the basic concepts required to understand managed realignment, including underlying drivers, terminology and methods of implementation. In Chapter 1, I explain the socio-economic and environmental drivers underpinning the need for managed realignment. The terminology associated with managed realignment is often used inconsistently in the literature and variable across countries. With that in mind in Chapter 2, I clarify the most common terminology and propose a new (and broader) definition for managed realignment to incorporate the different methods of implementation that exist. These different methods of implementation are explained in Chapter 3.

The following six chapters describe and illustrate the different managed realignment approaches and strategies implemented by different countries. Chapter 4 describes some of the relevant national and transnational strategies that are currently in place to support managed realignment and other innovative alternatives to manage coastal flood and erosion risk. Chapters 5, 6 and 7 illustrate the range of approaches that can be adopted to implement manage realignment in practice, varying from a combination of hard and soft engineering used in the Netherlands (Chap. 5), the removal of coastal defences used by the National Trust in England and Wales (Chap. 6) and managed retreat used in Maui, Hawaii (Chap. 7).

In Chap. 5, Joost Stronkhorst and Jan Mulder describe the coastal management strategies implemented in the Netherlands and provide examples of different managed realignment methods implemented along the sandy coast of the North Sea and the silty shorelines of estuaries and the Wadden Sea. It is common knowledge that the Netherlands is a low-lying country where land reclamation and protection against floods are of paramount importance. Therefore, the Dutch experience demonstrates that managed realignment, alongside other coastal engineering approaches, can be strategically implemented to provide coastal protection and promote sustainable uses.

In Chapter 6, Phil Dyke and Tony Flux describe the principles of coastal management adopted by the National Trust and present their experience in a project where managed realignment was implemented through removal of coastal defences in Brownsea Island, UK. The National Trust is a non-governmental charity organisation that owns about 10% of the coastline of England and Wales and is recognised worldwide as a role model of institutional capacity for promoting the sustainable management of coastlines. Chapter 7 provides an example of managed retreat alternatives for developed coastlines reliant on beach-related tourism. In this chapter Thorne Abbott describes how the implementation of a setback line in Maui (Hawaii, USA) assists managed retreat and contributes to maintain and enhance the local economy.

Chapters 8 and 9 focus on the UK, where managed realignment is undertaken mainly through the realignment of defences with the objective of creating intertidal habitat. These chapters benchmark the current state of affairs of managed realignment in the UK through the perspective of practitioners in the public and private sectors. In Chapter 8, Karen Thomas describes emerging policies and the drivers underpinning the implementation of managed realignment. She also examines the lessons learned so far by the Environment Agency, the government agency responsible for overseeing coastal flooding and erosion risk management in England and Wales. Chapter 9 brings the experience and views of Nigel Pontee, a private consultant involved in planning, designing and delivering managed realignment projects. He identifies and discusses the factors influencing the long-term sustainability of managed realignment.

Chapter 10 draws from existing literature and the results of a survey conducted in 2013 to expand on the challenges and achievements discussed in Chapters 8 and 9 to the wider context. This chapter discusses evidence of achievements attributed to managed realignment in relation to improvement of flood risk management, creation of intertidal habitats and the potential to deliver other ecosystem services such as carbon sequestration. It also compares how researchers, practitioners and stakeholders perceive managed realignment in the UK and elsewhere in the world.

Finally, Chapter 11 synthesises the current research and understanding presented in previous chapters to provide final remarks about the long-term sustainability of managed realignment. The Appendix provides a 'work in progress' list of managed realignment projects implemented in Europe and few examples of projects in the USA that fit the broad definition of managed realignment suggested in this book. Readers can keep updated with the progress of compiling a more comprehensive list of managed realignment projects, by accessing the online map of all identified projects through the link provided in the Appendix.

Managed realignment involves complex issues which vary in space and time. The knowledge of how the many social, economic and technical aspects of this strategy interact is evolving fast as new policies are formulated, more projects are implemented and new monitoring data become available. I have learned a great deal during the process of writing this book and hopefully you will find in these pages enough content to get you started in this interesting and relevant topic. Throughout the book you will find relevant sources of information you can access online to find out more and keep track of recent developments. I hope you enjoy reading.

February 2014   Luciana S. Esteves
Winchester, UK

Acknowledgements

I would like to thank all that directly or indirectly helped me accomplish the objective of writing the first book to be published on managed realignment. First of all I would like to thank Sherestha Saini (Acquisitions Editor for Environmental Sciences at Springer) for believing in the relevance of the topic and in my ability to develop this project (not to mention for understanding my many request for deadline extensions). Without her interest this book would still be just an idea. Thanks Sher!

I am very grateful to all contributors (named above) for accepting the invitation to write a book chapter providing their valuable practical experience on the many facets of managed realignment. Their state-of-the-art knowledge and relevant examples bring to the book a much needed balance between practice and academic research. An appreciation of support is given to over the 250 respondents of my on-line survey on managed realignment, part of the results are presented in Chapter 10.

I would also offer thanks for invaluable support from all who so kindly allowed their photographic or graphic work to be reproduced in this book or contributed with their knowledge and experience (listed here in surname alphabetical order): John Akerman for the beautiful aerial photos of Medmery (UK); Tony Baker (RSBP) for allowing the use of the Hesketh Marsh photo (UK); Oliver Beauchard (Netherlands Institute for Sea Research, NIOZ) for the photo of Lippenbroek and diagram of CRT; Pieter de Boer (Rijkswaterstaat, Netherlands) for assisting in the production of Chap. 5; Steve Burdett and Nick Cooper (Royal HaskoningDHV, UK) for the photo of the seawall realignment at South Tyneside; Olivia Burgess (Crown Estate, UK) for information on ownership of intertidal land in the UK; Andy Coburn (and Bill Neal and Norma Longo) for many fantastic photos; David Denton for his photos of the stormy Dawlish shoreline; Chris Fox (Wood Pawcatuck Watershed Association, USA) for the photos of the dam removal; Katharina Jacob (IBA Hamburg) for the photo and information about the Kreetsand project; Baerbel Koppe (AQUADOT, Germany) for the photo of the house rebuilt in the flood plain of the river Oder; Sylvain Moreira (DIMER/Groupe Environnement, France) for digging the photos of the Polder de Sebastopol (and permission of use); Tara Owens (County of Maui and University of Hawaii) for the photos of Charley Young's beach in Maui; Hadassah Schloss (Texas General Land Office, USA) for the photos and information about the relocation of houses in Texas; Peter Smith (http://www.petersmith.com/)

for his aerial photo of Alkborough (UK); John Tabernham (Environment Agency) for giving permission to use the aerial images of Steart (UK); Lucie Thiebot (Artelia Maritime, France) for providing information about managed realignment in France.

Many thanks to Jon Williams for peer-reviewing earlier drafts and keeping me on track; his comments greatly help improving content, structure and readability of the text. Last but not less important, I appreciate the contribution of Ayden Liddell and Samuel Lowther in helping to compile information for the book as their work placement in the summer of 2013. Ayden produced a GIS database of realignment projects in the UK and Sam looked into the effects of managed realignment on birds.

Finally, I would like to apologise to all family, friends and colleagues from Bournemouth University for my impatience and lack of time during the months I was working on this book.

Contents

About the Authors

Thorne Abbott, Coastal Resource Planner, Coastal Planners, Hawaii, USA

Thorne Abbott is currently a consultant specialised in assisting oceanfront land-owners in addressing erosion crisis and shoreline change in Hawaii. He has over 20 years of experience in natural resources management in the public and private sectors in continental USA and in the territories in the Pacific Ocean.

Phil Dyke, Coast and Marine National Specialist, The National Trust, Swindon, UK

Phil Dyke has 30 years of experience working for the National Trust on coastal and marine issues. In his current position since 2007, he supports the National Trust in England, Wales and Northern Ireland to implement sustainable coastal and marine management approaches. In addition to his role at the National Trust, Phil teaches coastal and marine management and he is the Chair of the EUROPARC Atlantic Isles Coast and Marine Working Group, which seeks to promote integrated management approaches in protected areas.

Luciana S. Esteves, Senior Lecturer in Physical Geography, Bournemouth University, Poole, UK

Luciana S. Esteves has over 20 years of experience in teaching and researching coastal processes and management in Brazil, the USA and the UK. Her research focuses on impacts of coastal protection, extreme events and climate change on flood and erosion risk. She has over 40 publications on these topics and is a member of the Natural Environmental Research Council Peer Review College and the editorial board of the Journal of Coastal Research.

Tony Flux, South West Region Coast and Marine Adviser, The National Trust, Morecombelake, UK

Tony Flux has extensive experience in issues concerning coastal management in South West England. As a Coastal Adviser for the National Trust since 2007, he supports projects along the Dorset and East Devon coasts. Previously, Tony worked for the Dorset Coast Forum and the Dorset Areas of Outstanding Natural Beauty team.

Jan P. M. Mulder, Senior Coastal Specialist, Deltares, Netherlands

Jan P. M. Mulder has over 25 years of experience in researching topics related to coastal morphology and management. He has worked at TNO, Dutch Ministry of Infrastructure & Environment (Rijkswaterstaat) and, since 2008, at Deltares. Jan is also part-time assistant professor at Twente University. His expertise includes translation of research into policy and the integration of field observations and empirical and numerical modelling concerning physical and ecological disciplines.

Nigel Pontee, Global Technology Leader Coastal Planning & Engineering, CH2M HILL, Swindon, UK

Nigel Pontee has 20 years of experience in consultancy work concerning coastal geomorphology and management, during which he has contributed to over 190 estuary projects, including the appraisal of over 150 habitat restoration sites and the design of managed realignment schemes in the UK. He has authored over 200 consultancy reports and over 75 publications on coastal and estuarine shoreline management aspects, including managed realignment and habitat creation design.

Joost Stronkhorst, Integrated Coastal Zone Management Senior Advisor, Deltares, Delft, Netherlands

Joost Stronkhorst has more than 30 years of experience in the development of water and sediment management in the Netherlands and multidisciplinary coastal research. He has worked for the Dutch Ministry of Infrastructure & Environment (Rijkswaterstaat) where he was involved in the Delta works and marine pollution control.

Karen Thomas, Senior Coastal Advisor (Eastern Area), Environment Agency, Ipswich, UK

Karen Thomas is a technical expert in flood risk management with 20 years of experience in management of coastal issues at the national and local levels. Karen has led the Essex and South Suffolk Shoreline Management Plan and works on managed realignment projects since joining the Environment Agency in 1999. She works with communities and stakeholders to encourage greater interest in managing the coast in a more sustainable manner.

List of Figures

List of Tables

Chapter 1
The Need for Adaptation in Coastal Protection: Shifting from Hard Engineering to Managed Realignment

Luciana S. Esteves

Abstract Climate change is already affecting our lives. Scientific predictions influence government policies; these in turn affect the way we live as individuals and society. The way we live is greatly dependent on the natural environment. The ultimate consequence of climate change to society is that we cannot continue living the way we do because the environment around us is changing. Therefore, we are all compelled to adapt to the new conditions and become more resilient to change as individuals and communities. Climate change and environmental and financial concerns have led to a shift from the traditional 'hold-the-line' approach of coastal protection towards more flexible soft engineering options. Managed realignment is a relatively new soft engineering approach aiming to maximise environmental and socio-economic benefits by creating space for coastal habitats to develop. The natural adaptive capacity of coastal habitats (and the ecosystem services they provide) underpins the concept of managed realignment. This chapter describes the main drivers leading to the implementation of managed realignment and the multiple functions it is expected to provide.

Climate change has become omnipresent in the media and scientific debates in recent decades. Irrespectively of whether you believe it is a natural or human-induced process, climate change is already affecting our lives directly or indirectly. Scientific predictions influence government policies; these in turn affect society behaviour locally, nationally and internationally. Policies underpinned by climate change concerns are reflected in many aspects of our lives, from the price, availability and provenience of services and goods (e.g. energy, water and food) to the way natural resources are managed and valued. The way we live is greatly dependent on the natural environment, which is changing in response to the new climatic conditions.

Directly through environmental changes or indirectly through government policies, climate change has consequences on management of waste, soil and water; urban planning; flood and erosion risk etc. The ultimate consequence of climate change to society is that we cannot continue living the way we do because the envi-

L. S. Esteves (✉)
Faculty of Science and Technology, Bournemouth University, Talbot Campus,
Poole, Dorset, BH12 5BB, UK
e-mail: lesteves@bournemouth.ac.uk

L. S. Esteves, *Managed Realignment: A Viable Long-Term Coastal Management Strategy?*,
SpringerBriefs in Environmental Science, DOI 10.1007/978-94-017-9029-1_1,
© Springer Science+Business Media Dordrecht 2014

Fig. 1.1 Adaptation can take many forms. This house had just been built (in Ziltendorfer Niede-rung, the floodplain of the River Oder in Germany) when it was flooded in the summer of 1997. Afterwards the owner decided to re-build the house in a slightly retreated position and on an elevated level, so the 'ground floor' would be less prone to flooding. The photo shows the original position of the house and how it looks after 'redevelopment' (photo courtesy of Baerbel Koppe, AQUADOT Engineering Consultants, Hamburg)

ronment around us is changing. Therefore, we are all compelled to adapt to the new conditions and become more resilient as individuals and communities (Fig. 1.1). Adaptation can be defined as "*adjustments in ecological-social-economic systems in response to actual or expected climatic stimuli, their effects or impacts*" (Smit et al. 2000, p. 225).

Predictions of climate change impacts in coastal areas often include sea-level rise and more frequent and intense extreme weather events (e.g. IPCC 2013). Such future conditions challenge risk management in areas prone to coastal flooding and erosion (Fig. 1.2) leading to a change in policy direction (e.g. Parliamentary Of-fice of Science and Technology 2009). Higher water levels and enhanced stormi-ness reduce the level of protection offered by existing coastal defences and increase maintenance costs.

At many coastal locations worldwide, upgrading of hard engineering defences is now constrained by both high economic costs and undesired environmental im-pacts. Climate change and environmental and financial concerns have led to a shift from the traditional 'hold-the-line' approach of coastal protection towards more flexible soft engineering options. In this context, managed realignment is an impor-tant adaptation measure aiming to improve the sustainability of coastal erosion and flood risk management in light of climate change.

Managed realignment is a relatively new soft engineering approach aiming to maximise environmental and socio-economic benefits by creating space for coastal habitats to develop. The natural adaptive capacity of coastal habitats (i.e. the abil-ity to dynamically adjust to changing environmental conditions) and the ecosystem

Fig. 1.2 A long spell of stormy weather has caused major impact at the coast of southwest England in January and February 2014. Coastal defences have been overwhelmed at Dawlish where the railway line collapsed cutting rail links to Cornwall and west Devon. A number of houses located next to the rail line were affected and the seawall and groynes were greatly damaged

services[1] they provide (i.e. the benefits society gain from the presence of functional ecosystems) are key to the concept of managed realignment. This chapter describes the main drivers leading to the implementation of managed realignment and the multiple functions it is expected to provide.

1.1 The Need to Shift from Hold-the-Line to Managed Realignment

For centuries, hard engineering structures have been built to protect assets at the coast from erosion and flooding events. Hard engineering has controlled the dynamic nature of floodplains and coasts (Fig. 1.3) to safeguard human occupation and economic activities. As a result, many natural habitats have been destroyed and a large number of people and assets are now located in hazard-prone zones.

Coastal areas were once comprised of natural habitats that acted as a buffer between marine and terrestrial environments, offering natural protection against the dynamic nature of the sea. Floodplains, created by river dynamics to accommodate the excess of water in times of increased rainfall, are now occupied by housing and industry and turned into impermeable land with limited capacity to take up overflown waters. The impact of engineering on river dynamics has affected the coast in many ways, notably by reducing the supply of sediments, affecting the ability of natural environments to dynamically adjust to changing environmental conditions and leading to the degradation or loss of natural habitats.

These hard structures have created a legacy of coastal management problems, which are now considered unacceptable. To avoid such detrimental effects, construction of new hard engineering structures have been banned or strictly controlled in some coastlines (e.g. North Carolina's Coastal Area Management Act, *§113A-115.1 Limitations on erosion control structures*). While the nineteenth century and part of the twentieth century marked the era of hard engineering works reshaping and altering the natural balance of coastal systems, throughout the twentieth century soft engineering has risen as a preferred alternative to reduce the detrimental impacts caused by artificially fixed coastlines.

Soft engineering measures 'work with natural processes' benefitting from coastlines that are able to evolve more dynamically. Underpinning the implementation of soft engineering is the need to restore the capacity of coastal environments to carry out their natural coastal protection function and deliver other ecosystem services. Beach nourishment, dune restoration and more recently managed realignment are examples of soft engineering schemes (Fig. 1.4) that have been increasingly implemented worldwide.

The shift away from hard engineering does not stop by the increasing popularity of soft engineering; de-engineering efforts (i.e. the deliberate removal of engineer-

---

[1] The Millennium Ecosystem Assessment (2005) was the benchmark for the assessment of ecosystem services worldwide and the consequences of ecosystem changes to society. All technical and synthesis reports are available from: www.maweb.org.

Fig. 1.3 Hard engineering structures are shaping coasts and rivers worldwide, causing natural habitat loss and favouring development in hazard-prone areas. (**a**) Storms have greatly reduced the beach fronting coastal defences in Selsey, an area also prone to flooding (photo: L.S. Esteves). (**b**) Flood defence along the Mississippi river close to where a breach occurred during hurricane Katrina causing devastating flooding (photo: L.S. Esteves). (**c**) Cancún is a narrow barrier beach densely occupied by hotels and subjected to intense erosion due to hurricanes impacts; hotels survival depends on beach nourishment and hard engineering (photo courtesy of Grupo de Ingeniería de Costas y Puertos del Instituto de Ingeniería, UNAM). (**d**) Map Ta Phut suffers critical coastal problems due to the engineering works that support the largest industrial park in the country (photo courtesy of Andy Coburn, Western Carolina University)

Fig. 1.4 Examples of soft engineering schemes: (**a**) beach nourishment in Palm Beach, Florida, USA (photo: L.S. Esteves); (**b**) dune restoration in Lido de Sete, French Mediterranean coast (photo: Jon J. Williams); and (**c**) managed realignment in Medmerry, UK (photo: John Akerman)

Fig. 1.5 (**a**) Lower Shannock Falls Dam Removal and Fish Passage project was conducted as part of a wider strategy to restore natural conditions at the Pawcatuck Watershed Start (photo by Chris J. Fox, Wood Pawcatuck Watershed Association, Rhode Island). (**b**) Elwha river dam in the state of Washington after removal. (**c**) Dillsboro Dam on the Tuckaseegee River in North Carolina being removed (photos b and c: Andy Coburn, Western Carolina University)

ing structures) are becoming more common. In the twenty-first century, innovative de-engineering projects are being implemented in an effort to restore the environment to a more natural functioning. Examples of such de-engineering efforts include: removal of river dams in the USA (Fig. 1.5) and the removal of coastal defences (see Chap. 6).

River restoration projects including removal of dams are becoming more common in the USA to restore salmon migration and sediment supply to the coast. According to *American Rivers* 51 dams were removed from U.S. rivers in 2013 adding to the almost 850 removed in the last 20 years. Dams retain sediment and prevent them reaching the coast causing a sediment deficit that can lead to coastal erosion. The largest of such river restoration projects is being conducted in the Elwha River in Washington, USA. The Elwha Dam was completely removed in 2012 (Fig. 1.5b) and the removal of the Gilnes Canyon Dam is expected to be completed in September 2014.[2] Over 100 years these dams retained an estimated sediment load of 26 million m³. Preliminary results indicate that about 18 % of this sediment storage have already been mobilised and are slowly reaching the Strait of Juan de

---

[2] Videos, photos and information about the Elwha River dam removal project can be found at: http://www.nps.gov/olym/naturescience/elwha-ecosystem-restoration.htm.

Fuca coast, which received about 1.2 million m^3 of new sediment during the winter 2012–2013 (MacDonald and Harris 2013).

Only time will tell whether de-engineering will become widespread in the twenty-first century as the ultimate solution to current coastal management problems. In the meanwhile, soft engineering methods are currently the most common response to the environmental and economic problems created by hard engineering. Managed realignment, in particular, is appearing as a sustainable option to create opportunities for the development of habitats able to provide desired multiple functions, including: (1) compensation or offsetting intertidal habitat loss due to coastal squeeze and developmental pressures; (2) sustainable coastal protection by dissipating wave energy and/or acting as flood-water storage areas; and (3) other ecosystem services such as carbon sequestration, amenity value etc.

There are various interpretations about what types of mechanisms or methods of implementation can be considered as managed realignment and a clarification of terms, including a new definition is proposed in Chap. 2. Schemes described as managed realignment are most commonly implemented in low-lying estuarine or open coast sites and often include breaching or removal of existing defences (Fig. 1.4c). In many sites, the construction of a new line of defences further inland is required to control flood risk. Hence the expression 'managed realignment' may refer to the inland relocation of both the coastline and the flood defence line.

1.2   Compensation or Offsetting of Coastal Habitats Loss

Only recently, the realisation about the cascading environmental impacts caused by engineering works translated into restoration actions. A good part of coastal habitats worldwide have been lost or degraded as a consequence of human activities (e.g. Gedan et al. 2009). It is estimated that only about 50 % of the saltmarshes remain worldwide (Barbier et al. 2011). A recent assessment (Dahl and Stedman 2013) indicates that about 50 % of wetlands in the continental USA have been lost, with coastal wetlands being lost at rates of 32,300 ha per year on average from 2004–2009, reflecting a 25 % increase over the previous 6 years. Similar losses of coastal wetlands are found in many countries worldwide, e.g. 51 % in China (An et al. 2007); 70 % in Singapore (Yee et al. 2010).

Land reclamation and coastal development are two major causes for the reduction in the extent of intertidal habitats. Recognitions of the importance of coastal habitats to society have led to the creation of legislation aiming to prevent further human-induced habitat loss and, through habitat creation, compensate unpreventable losses and offset past impacts.

Environmental regulations are now enforced in many locations to mitigate further damage and to compensate for loss of natural habitats (e.g. van Loon-Steensma and Vellinga 2013). European legislation (especially the Habitats Directive and the

Birds Directive[3]) has been a fundamental driver for nature conservation efforts in the countries member of the European Union (EU). To protect habitats and species of European importance, an EU-wide network of designated conservation sites (called Natura 2000) has been established. Each EU country is obliged to transpose the EU Directives into national legislation. National governments are responsible for identifying which geographical locations must be designated Natura 2000 sites[4] within their territory and for taking all necessary measures to ensure the protection of designated habitats and species.

Many coastal habitats in Europe, including most intertidal flats and saltmarshes, are now within Natura 2000 sites. Similar transnational efforts are in place elsewhere in the world. For example, the North American Waterfowl Management Plan was established in 1986 to address the decline in waterfowl populations associated with the destruction of wetlands in Canada and the USA. Mexico joined the Plan in 1993. However, it is at the national legislation that compensation for loss of natural habitats is most commonly addressed, e.g. the USA *National Ocean Policy Implementation Plan* (National Ocean Council 2013); Australia's *Environment Protection and Biodiversity Conservation Act 1999*; Canada's Federal Policy on Wetland Conservation of 1991; *New Zealand's Resource Management Act 1991* to name just a few.

Most often, these legislations establish that human-induced loss or damage to protected coastal habitats (e.g. by the constructions of ports and marinas, dredging activities, coastal defence works etc.) must be prevented, but if inevitable, it must be compensated by recreation or restoration of equivalent habitats. In Europe, it is also legally binding to offset long-term habitat loss due to coastal squeeze.

1.2.1 Coastal Squeeze

Coastal squeeze and land reclamation are often cited as the main causes for the loss of intertidal habitats (e.g. Doody 2013). The term 'coastal squeeze' is most commonly used in the literature referring to the loss of intertidal habitats caused by rising sea levels along coastlines fixed by hard engineering structures (e.g. French 2004; Pontee 2013). Coastal squeeze reflects loss of habitats caused by the presence of hard engineering structures and should not refer to losses due to natural processes (Pontee 2013).

---

[3] Over 200 habitat types and 1,000 species of plants and animals of European importance are protected under the Habitats and Birds Directives. More information about the European nature and biodiversity legislation is found at: http://ec.europa.eu/environment/nature/legislation/habitatsdirective/.

[4] Two types of designated areas form part of the Natura 2000 network: Special Areas of Conservation (SAC) and Special Protection Areas (SPA). SACs are designated for habitat conservation under the Habitats Directive. SPAs are designated under the Birds Directive for supporting species or populations of birds of European importance. The distribution and location of the designated Natura 2000 sites can be viewed at: http://natura2000.eea.europa.eu/.

Natural coasts will dynamically change to adjust to new meteorological and oceanographic conditions, including short- and long-term sea-level fluctuations. In natural systems, rising sea levels usually result in a landward migration of habitats (Fig. 1.6a, b), a process characteristic of transgressive coasts. Depending on a number of interacting physical and biotic variables (e.g. sediment availability, rate of sea-level rise, coastal topography, presence and type of vegetation), habitats such as saltmarshes are able to migrate inland and accrete vertically without resulting in loss of intertidal habitat. The extent of intertidal area and the type of habitat that will develop depend on the coastal topography and whether both low and high water lines are able to move inland at similar rates.

The type of intertidal wetland that may be established at any particular location is influenced (amongst other variables) by their position within the tidal range (Fig. 1.6a). The vertical zonation of marshes reflects the tolerance of species to inundation (Pennings and Calloway 1992), i.e. more tolerant species are found at lower elevations (which are flooded more often). Saltmarshes tend to form between the mean high neap and mean high spring water levels; pioneer species colonise areas between mean low neap and mean high neap water levels and mudflats develop between mean low spring and mean low neap water levels.

Coastal defences fix the landward boundary of intertidal habitats by preventing the high water line to move further inland (Fig. 1.6c). Therefore, a rise in sea level will gradually increase the frequency and duration of inundation and ultimately result in loss of intertidal area as lower areas become permanently submerged (Fig. 1.6d). Depending on the range of elevations in relation to the water levels, increased exposure to inundation may lead to a shift in the types of marsh communities and/or the loss of habitats. Mudflats may occupy areas formerly dominated by pioneer marshes; these might shift to higher ground and eventually species less tolerant to inundation will disappear if suitable conditions are not available (Fig. 1.6d).

Considering that some countries are legally bound to compensate for habitat lost due to human-induced processes, it is important to identify when and where loss and degradation is due to natural causes. Storm impacts for example are partly responsible for the loss of wetlands along the USA Gulf of Mexico coastline (Dahl and Stedman 2013). Hughes and Paramor (2004) suggest that increases in the abundance of the polychaete *Nereis* might be the cause of widespread loss of pioneer marshes in south-east England. However, it is widely accepted that anthropogenic activities are the major causes of intertidal habitat loss and legislation efforts now aim to offset this negative impact.

Many coastlines in Europe are protected by hard engineering defences and many of these structures are very close or at the boundaries of Natura 2000 sites. With the continuation of rising sea levels, EU countries are required to compensate for coastal squeeze likely to affect protected sites throughout the life-time of the hard engineering structures. Therefore, not only the construction of new coastal defences but also the upgrading or even the maintenance of existing structures must be carefully assessed. For example, policy guidance in the UK (Defra Flood Management Division 2005) indicates that any project negatively affecting a Natura 2000 site can only be allowed if all the following three criteria apply: (1) no other alternatives

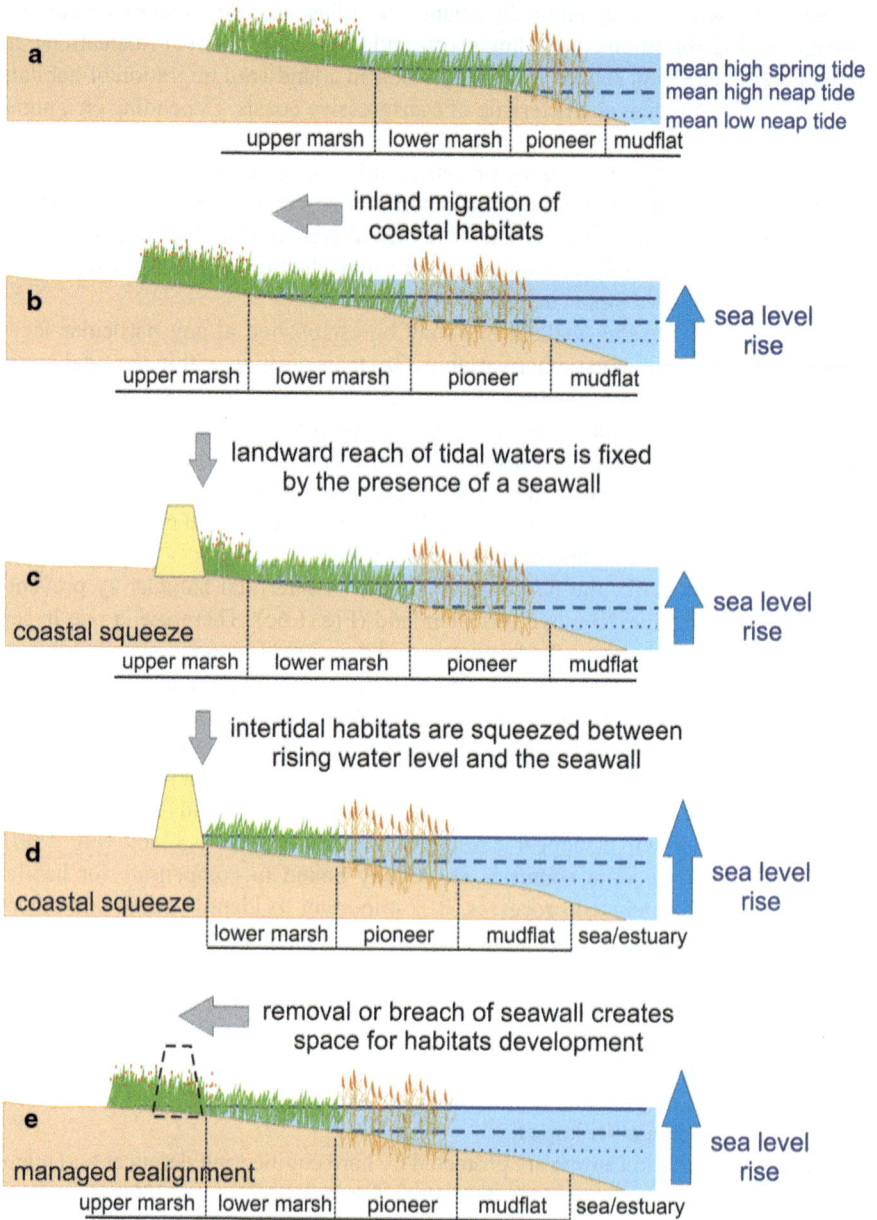

Fig. 1.6 Coastal habitats develop as a function of interacting physical and biological characteristics. (**a**) The elevation in relation to the tidal range is one of the key factors determining the type of intertidal habitat. (**b**) As a response to rising sea levels, coastal habitats tend to migrate inland and the intertidal area may expand or reduce depending, for example, on the coastal topography. (**c**) Hard engineering structures will invariably fix the landward limit of intertidal areas, which will be reduced in extent as sea levels rise and more land becomes permanently inundated. (**d**) The loss of coastal habitats due to rising sea levels in front of artificially fixed shorelines is known as coastal 'squeeze'. (**e**) Managed realignment is often implemented through planned breaching or removal of coastal defences to create space for the development of intertidal habitats with the aim of improving flood risk management with added environmental value

exist; (2) the project is necessary for imperative reasons of overriding public inter-est; and (3) measures are in place to ensure that ecological functions lost will be compensated so the coherence of the Natura 2000 network is maintained.

Managed realignment most often involves planned breaching or removal of coastal defences. By allowing tidal waters to flow further inland (Fig. 1.6e), new intertidal areas and accommodation space for sediment deposition are created. The development of intertidal habitats enhances local biodiversity and the sustainability of coastal protection and, therefore, is crucial for the success of managed realign-ment as a sustainable coastal management approach. Managed realignment is play-ing an important role in providing the opportunities for the creation of habitats required by law. Often, countries adopt an estuary-wide or regional strategy (see Chap. 4) and evaluate in an integrated manner the extent in area and the type of habitats that will be lost within the physiographic region and the most suitable sites where compensatory measures might be implemented.

1.3   Sustainable Flood and Erosion Risk Management

Sea-level rise and intense extreme weather events threaten the sustainability of coastal defences worldwide. The challenges are not only the increasing financial costs to maintain coastal defences and provide the required level of protection against coastal flooding and erosion; they also relate to the loss of intertidal habitat and the wider consequences to the environment, the economy and people's well-be-ing. In light of climate change and current environmental legislation, governments and private land owners have to adopt strategic approaches that lead to improved long-term financial, environmental and social sustainability. Generally, 'improved sustainability' of flood and erosion risk management is translated into the following needs:

a. *To reduce the costs of maintaining coastal defences*. There are two ways of cut-ting such financial costs: by shortening the overall length of defences and/or the fre-quency in which repairs are required. Both hard engineering and managed realign-ment can be used to reduce the length of the shoreline that needs to be protected. In the Netherlands, hard engineering has greatly shortened the country's shoreline length (see Chap. 5). Managed realignment can be designed to remove defences in sections of the shoreline or reduce the overall length through realignment (Fig. 1.7).

In the UK, 53 managed realignment schemes have been implemented (see a list in the Appendix), in 35 sites through removal, breaching or realignment of de-fences. A new defence line was built further inland in 20 sites; no new line of de-fence was built in 11 sites; no information was found for the remaining four sites (Esteves 2013). New defences were not required in some sites because a secondary line of defence was already in place (and it was upgraded as part of the realignment scheme) or there was no flood risk to inland areas (e.g. due to higher topography). In 17 sites, the total length of the new defences was similar or considerably longer

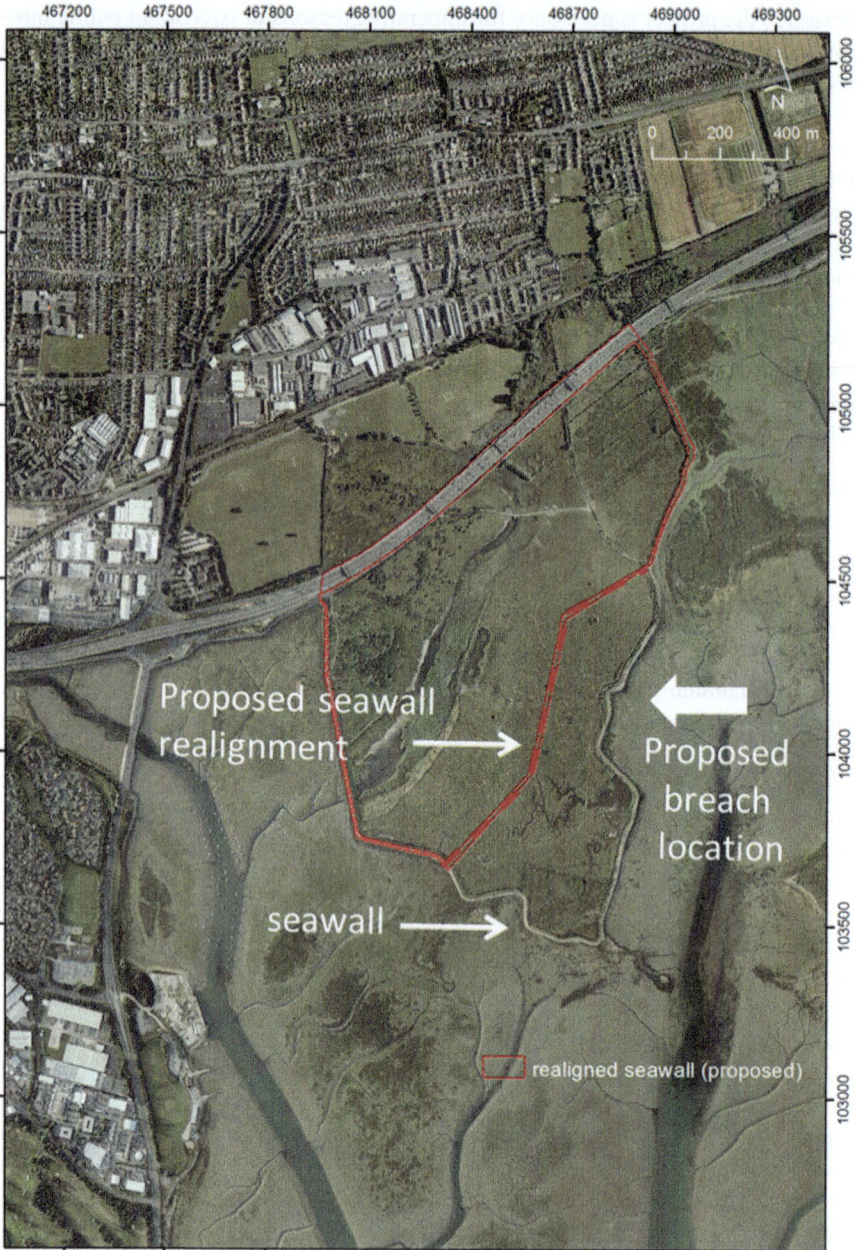

Fig. 1.7 Farlington Marshes (Portsmouth, southern England) was reclaimed in the 1770s and is enclosed by sea defences. Managed realignment to restore tidal flow into about 30 ha has been suggested as a preferred coastal management option to be implemented in the medium term. Plans include breaching of the existing seawall and realign the defences as indicated here. This realignment would reduce the length of the defences to be maintained by about 700 m. Additionally, less frequent repairs would be required if saltmarshes develop within the site helping dissipation of wave energy in front of the new defence line (Aerial image dated 21 July 2008, available from the Channel Coastal Observatory, www.channelcoast.org)

than the old defence line (length ratio varied from 70–190%). Only in three sites, the length of new defences is considerably shorter than the old defences.

It is expected that the realignment site will act as a sink for sediments, favouring the development of saltmarshes. The resulting wider intertidal profile provides natural coastal protection through dissipation of wave energy (French 2004; Shepard et al. 2011; Spalding et al. 2013), which tends to be significantly greater over saltmarshes than over un-vegetated intertidal flats (Möller and Spencer 2002). The UK National Ecosystem Assessment (2011) estimates that the first 10–20 m of saltmarshes are able to attenuate about 50% of the incoming wave energy. The same report indicates that coastal habitats provide to England £ 3.1–33.2 billion[5] savings in costs of engineering defences and the natural defence offered by sand dunes in Wales alone was valued to be worth between £ 53 and £ 199 million in 2007.

Saltmarshes, beaches and dunes help to protect the existing line of defences they front, reducing the frequency of maintenance and the size of defences required. Managed realignment projects often include cost-benefit analysis in the planning stage. However, studies validating the original estimates or presenting an assessment of the maintenance cost-savings accruing from the creation of coastal habitats in the realigned sites are rare or not readily available. All managed realignment projects in the UK have been implemented at sites where flood defences were in poor conditions. Considering that these defences were not being maintained and that many sites have now a longer line of defence, it is difficult to ascertain whether managed realignment has resulted in lower maintenance costs.

b. *To reduce risk to people and key assets*. This need can be addressed in two (somewhat conflicting) approaches: by improving the level of protection provided to hazard-prone areas or by reducing the number of people and assets located in these areas. Improving the level of protection offered to certain areas might result in increased number of people and property at risk if no planning regulation is in place. Areas with new or improved defences are often perceived as 'safe' and become attractive for dwellings and businesses. The issue here is that coastal protection schemes are designed to offer protection against events of certain magnitudes and cannot prevent impacts from events exceeding in magnitude, duration and or combination of factors not anticipated in the design.

The consequences of such 'extreme' events can be catastrophic as illustrated by the aftermath of the stormy weather in the 2013–2014 winter causing widespread coastal erosion and flooding in the UK; hurricane Sandy in 2012 along the coasts of New York and Long Island (USA); cyclone Xynthia in 2010 affecting the French coast along the Bay of Biscay and hurricane Katrina in 2005 along the embankments of the Mississippi River in New Orleans (USA). Although improving the level of protection offered by coastal defences might be desired in populated areas, the only safe climate-proof response at all temporal scales is to reduce the number of people and assets in high risk areas.

---

[5] The range shown here is the actual figure given in the cited document—it is not a typographical error.

The level of protection can be improved by making defences more robust or more efficient. Environmental legislation might prevent upgrading or building hard defences; therefore efforts tend to resource on soft engineering alternatives, including managed realignment. The advantage of soft engineering resides on recreating coastal habitats to benefit from their capacity to provide natural protection against storms. Wide dune and beach systems, mangroves and saltmarshes act as a buffer, reducing the water level and wave energy reaching areas further inland. Therefore, this natural storm buffering capacity can improve the level of protection offered to people and property (Spalding et al. 2013). However, it is important to recognise that the ability to offer protection might be limited in habitats at early stages of development.

Although many managed realignment projects are based around the capacity of coastal habitats to offer natural protection during storms, the effects of habitat creation in realigned sites are still to be quantified. Despite improved flood risk management being a major drive for managed realignment, there are virtually no studies addressing how land use changes on realigned sites have affected flood risk to adjacent areas. Recent studies indicate that breaching of defences can cause significant hydrodynamic changes to the local coastal system (Friess et al. 2014) and it is important to understand the potential consequences to flood risk.

c. *To ensure no preventable damage or loss of natural habitats.* As discussed previously, legislation may be in place limiting coastal defence works if they are likely to cause damage or loss of natural habitats; and compensation is required for impacts on designated conservation areas. Such requirements pose a challenge to flood and erosion risk control, as in some places either removing existing defences or upgrading them will cause harm to one type of habitat or another.

Farlington Marshes (Fig. 1.7), for example, was reclaimed over 200 years ago and the presence of seawalls allowed the development of freshwater habitats that are now protected by national and European legislation. The site supports a number of protected bird species and attracts bird watchers from far afield; it has also important recreational and amenity value for the local population. Managed realignment in the area will undoubtedly result in loss of the freshwater habitats affecting the site's capacity to support designated species. There is also potential for impact on flood risk to the area north of the site, including over 550 dwellings, a major road and the rail line. Managed realignment in Farlington Marshes would create intertidal habitat estimated to compensate for about 100 years of local loss due to coastal squeeze.

In case such as this, it is a matter of having to choose between maintaining the existing habitat (which took more than 200 years to develop and will need to be compensated if lost) or replacing it to create intertidal habitats to compensate for others types of habitats already lost. The issue here is then to decide between the uncertain gains brought by managed realignment; which will depend on the type and quality of intertidal habitats that might develop; and the certain loss of realised values; which often cannot be re-created locally or in the short-term (Esteves 2013). Similar situation is found in other locations in Europe and the conflict is well described by Maltby (2006, p. 93):

Table 1.1 Coastal ecosystem services, the benefits they bring to society, the importance of coastal areas in the overall provision of benefits and the potential for improvement through managed realignment

Ecosystem services[a]		Benefits to society	Importance[b]	Potential
Provision	Food provision	Fisheries, shellfish, algae	Medium-high	Undervalued
	Water storage and provision	Water availability	Medium-low	n/a
	Biotic materials and biofuels	Genetic resources, raw materials, medicinal resources, ornamental resources	Medium-low	n/a
Regulation	Water quality regulation	Sequestration of pollutants, water purification	High	Undervalued
	Air quality regulation	Sequestration of pollutants, gas regulation	Low	n/a
	Coastal protection	Natural hazard regulation, erosion and flood control, water and sediment flow regulation	High	High
	Climate and weather regulation	Carbon and nitrogen sequestration, temperature and humidity regulation	Medium-high	High
	Ocean nourishment	Nutrient cycling; soil formation	Medium-low [c]	Low
	Life cycle maintenance	Maintenance of genetic diversity, habitat	High	High
	Biological regulation	Pest and disease control	Low	Low
Cultural	Symbolic and aesthetic values	Cultural heritage and identity, well-being	High	Undervalued
	Recreation and tourism	Leisure and recreation	High	High
	Cognitive effects	Information, education, inspiration	High	Undervalued

n/a indicates ecosystem services that have not yet being associated to managed realignment
[a] Classification of ecosystem services modified from Liquete et al. (2013)
[b] Adapted from UK National Ecosystem Assessment (2005)
[c] As reported for soil quality-purification

We are then confronted by the contradictory situation of ecosystem destruction and re-establishment both featuring prominently in society's agenda. The challenge is to manage the processes of change so that we do not irretrievably lose assets difficult or impossible to replace.

1.4   Delivery of Other Ecosystem Services

The great interest in the implementation of managed realignment is the potential to deliver multiple functions through the creation of habitats that are able to provide a range of ecosystem services (Table 1.1). The functions each ecosystem is able to execute depend on its biophysical characteristics (i.e. the interactions between biota and the physical environment). The ecosystem services include all the functions that contribute to human well-being and, therefore, have a socio-economic value (e.g. Haines-Young and Potschin 2010). Coastal ecosystems provide more services contributing to human well-being than other ecosystems (Millennium Ecosystem Assessment 2005). Estimates at the global scale indicate that 77 % of the total economic value produced by the world's ecosystems are generated at the Coastal (Martínez et al. 2007).

Table 1.1 lists the services provided by coastal ecosystems and the benefits they offer to society. It also shows the importance of coastal habitats for the overall provision of the identified benefits; which is a qualitative evaluation based on the UK National Ecosystem Assessment (2011). The relative importance of coastal ecosystem services varies depending on the geographical characteristics and the spatial scale under consideration (e.g. higher importance may be expected in island nations than in large continental countries).

As managed realignment creates space for the development of costal habitats, it has the potential to enhance the provision of specific ecosystem services (Table 1.1). Different types of ecosystems might be more or less able to deliver each one of these services and associated benefits; therefore the potential for enhancement through managed realignment depends on the type of habitats that will be created. For example, coastal wetlands provide habitat for wildlife (e.g. spawning grounds, nurseries, shelter, and food for fish, shellfish, birds and other wildlife), help improve water quality by filtering runoff from agricultural and urban areas, and act as a buffer against storm and wave damage (e.g. Dahl and Stedman 2013).

Not all ecosystems services listed in Table 1.1 have been associated to 'potential' enhancement as a result of managed realignment (here identified as n/a); however, it does not mean that the potential does not exist. On the other hand, some of these ecosystem services are the main objective of realignment projects and therefore are identified as 'high' potential for enhancement. Other ecosystem services are rarely explicitly mentioned as a planned outcome of managed realignment, despite their high importance in coastal areas – these are indicated as 'undervalued' potential in Table 1.1.

Coastal protection and life cycle maintenance are the two primary ecosystem services sought to be enhanced through managed realignment. However, research studies so far have greatly focused on the biodiversity aspects, while little knowledge has been produced on the effects on flood and erosion regulation. Similarly, there is a high recognition for the potential of managed realignment to provide cultural ecosystem services but little progress has been made to quantify the benefits accruing from existing projects. Recently, more studies are focusing on biogeochemistry and the effects of land use changes in realigned sites on carbon and nitrogen flux. A brief review of current knowledge on the achievements of managed realignment on delivering creation of habitats and associated ecosystem services is provided in Chap. 10.

References

An, S., Li, H., Guan, B., Zhou, C., Wang, Z., Deng, Z., Zhi, Y., Liu, Y., Xu, C., Fang, S., Jiang, J., & Li, H. (2007). China's natural wetlands: Past problems, current status, and future challenges. *Ambio, 34*, 335–342.

Barbier, E. B., Hacker, S. D., Kennedy, C., Koch, E. W., Stier, A. C., & Silliman, B. R. (2011). The value of estuarine and coastal ecosystem services. *Ecological Monographs, 81*, 169–193.

Dahl, T. E., & Stedman, S. M. (2013). *Status and trends of wetlands in the coastal watersheds of the Conterminous United States 2004–2009* (p. 46). U.S. Department of the Interior, Fish and Wildlife Service and National Oceanic and Atmospheric Administration, National Marine Fisheries Service.

Defra Flood Management Division. (2005). *Coastal squeeze implications for flood management, the requirements of the European Birds and Habitats Directives*. Defra Policy Guidance. https://www.gov.uk/government/uploads/system/uploads/attachment_data/file/181444/coastalsqueeze.pdf. Accessed 3 Feb 2014.

Doody J. P. (2013). Coastal squeeze and managed realignment in southeast England, does it tell us anything about the future? *Ocean & Coastal Management, 79*, 34–41.

Esteves, L. S. (2013). Is managed realignment a sustainable long-term coastal management approach? *Journal of Coastal Research*, Special Issue 65(1), 933–938.

French, P. W. (2004). *Coastal and Estuarine Management*. London: Routledge.

Friess, D. A., Möller, I., Spencer, T., Smith, G. M., Thomson, A. G., & Hill, R. A. (2014). Coastal saltmarsh managed realignment drives rapid breach inlet and external creek evolution, Freiston Shore (UK). *Geomorphology, 208*, 22–33.

Gedan, K. B., Silliman, B. R., & Bertness, M. D. (2009). Centuries of human driven change in salt marsh ecosystems. *Annual Review of Marine Science, 1*, 117–141.

Haines-Young, R. H., & Potschin, M. P. (2010). The links between biodiversity, ecosystem services and human well-being. In D. G. Raffaelli & C. L. J. Frid (Eds.), *Ecosystem ecology: A new synthesis. Cambridge: BES ecological reviews series* (pp. . 110–139). Cambridge: Cambridge University Press.

Hughes, R. G., & Paramor, O. A. L. (2004). On the loss of saltmarshes in south-east England and methods for their restoration. *Journal of Applied Ecology, 41*, 440–448.

IPCC (Intergovernmental Panel on Climate Change). (2013). Summary for policymakers. In T. F. Stocker, D. Qin, G.-K. Plattner, M. Tignor, S. K. Allen, J. Boschung, A. Nauels, Y. Xia, V. Bex, & P. M. Midgley (Eds.), *Climate change 2013: The physical science basis. contribution of working group I to the fifth assessment report of the intergovernmental panel on climate change*. Cambridge: Cambridge University Press.

Liquete, C., Piroddi, C., Drakou, E. G., Gurney, L., Katsanevakis, S., Charef, A., & Egoh, B. (2013). Current status and future prospects for the assessment of marine and coastal ecosystem services: A systematic review. *PLoS One, 8*(7), e67737. doi:10.1371/journal.pone.0067737.

MacDonald, J. M., & Harris, N. (Ed.) (2013). Proceedings of the 8th Annual Elwha Nearshore Consortium Workshop (27 February 2013), Port Angeles, Washington. http://www.clallam.net/ccmrc/documents/ENC_2013_Proceedings.pdf. Accessed 10 Feb 2014.

Maltby, E. (2006). Wetland conservation and management: Questions for science and society in applying the ecosystem approach. In: R. Bobbink, B. Beltman, J. T. A. Verhoeven, & D. F. Whigham (Eds.), *Wetlands: Functioning, biodiversity conservation, and restoration*. Ecological Studies, 191(2), 93–116.

Martínez, M. L., Intralawan, A., Vázquez, G., Pérez-Maqueo, O., Sutton, P., Landgrave, R. (2007). The coasts of our world: Ecological, economic and social importance. *Ecological Economics, 63*, 254–272.

Millennium Ecosystem Assessment (2005). Ecosystems and human well-being: Wetlands and water—Synthesis. Washington: Island Press.

Möller, I. and Spencer, T. (2002) Wave dissipation over macro-tidal saltmarshes: Effects of marsh edge typology and vegetation change. *Journal of Coastal Research*, Special Issue 36, 506–521.

National Ocean Council 2013. National Ocean Policy Implementation Plan. http://www.
    whitehouse.gov/sites/default/files/national_ocean_policy_implementation_plan.pdf. Accessed
    1 Feb 2014.
Parliamentary Office of Science and Technology. (2009). Coastal management, *Postnote*, 342,
    p. 5. www.parliament.uk/briefing-papers/POST-PN-342.pdf. Accessed 7 Feb 2014.
Pennings, S. C., & Calloway, R. M. (1992). Salt marsh plant zonation: The relative importance of
    competition and physical factors. *Ecology, 73,* 681–690.
Pontee, N. (2013). Defining coastal squeeze: A discussion. *Ocean & Coastal Management, 84,*
    204–207.
Shepard, C. C., Crain, C. M., & Beck, M. W. (2011). The protective role of coastal marshes:
    A systematic review and meta-analysis. *PLoS One, 6*(11), e27374. doi:10.1371/journal.
    pone.0027374.
Smit, B., Burton, I., Klein, R. J. T., & Wandel, J. (2000). An anatomy of adaptation to climate
    change and variability. *Climatic Change, 45,* 223–251.
Spalding, M. D., McIvor, A. L., Beck, M. W., Koch, E. W., Möller, I., Reed, D. J., Rubinoff, P.,
    Thomas, S., Tolhurst, T. J., Wamsley, T. V., van Wesenbeeck, B. K., Wolanski, E., Woodroffe,
    C.D. (2013). Coastal ecosystems: A critical element of risk reduction. *Conservation Letters,
    00,* 1–9. doi: 10.1111/conl.12074.
UK National Ecosystem Assessment. (2011). *The UK national ecosystem assessment: Synthesis of
    the key findings*. Cambridge: UNEP–WCMC.
van Loon-Steensma, J. M., & Vellinga, P. (2013). Trade-offs between biodiversity and flood pro-
    tection services of coastal salt marshes. *Current Opinion in Environmental Sustainability, 5,*
    320–326.
Yee, A. T. K., Ang, W. F., Teo, S., Liew, S. C., & Tan, H. T. W. (2010). The present extent of man-
    grove forests in Singapore. *Nature in Singapore, 3,* 139–145.

Chapter 2
What is Managed Realignment?

Luciana S. Esteves

Abstract Many definitions of managed realignment exist. The understanding of what the term actually represents in practice has evolved through time and varies regionally, across sectors and among practitioners. A common understanding of managed realignment is further complicated by the use of other related terms; sometimes synonymous with managed realignment while at other times reflecting different concepts. Terms such as managed retreat, setback, regulated tidal exchange and depoldering have all being used in the literature associated with managed realignment, many times inconsistently. The lack of clarity in the use of terminology has contributed to negative connotations expressed by some stakeholders and the general public. This chapter clarifies the terminology currently in use and proposes a wider definition of managed realignment so it can be applicable to encompass the many forms of implementation adopted worldwide. Within this broader context, managed realignment becomes a general term that can be used to describe collectively the many mechanisms implemented to allow coastlines to evolve more flexibly with the objective of promoting more sustainable flood and erosion risk management.

2.1 Current Definitions

Many definitions of managed realignment exist. The understanding of what the term actually represents in practice has evolved through time and varies regionally, across sectors and among practitioners. A common understanding of the term is further complicated by the use of other related terms; sometimes synonymous with managed realignment while at other times reflecting different concepts. The description that more widely reflects the range of existing definitions (as it indicates different forms of implementation and the associated objectives) is provided below:

> Managed realignment means the deliberate process of realigning river, estuary and/or coastal defences. This may take the form of retreating to higher ground, constructing a setback line of defence, shortening the overall defence length to be maintained, reducing wall

L. S. Esteves (✉)
Faculty of Science and Technology, Bournemouth University, Talbot Campus,
Poole, Dorset, BH12 5BB, UK
e-mail: lesteves@bournemouth.ac.uk

L. S. Esteves, *Managed Realignment: A Viable Long-Term Coastal Management Strategy?*,   19
SpringerBriefs in Environmental Science, DOI 10.1007/978-94-017-9029-1_2,
© Springer Science+Business Media Dordrecht 2014

or embankment heights or widening a river flood plain. The purpose of managed realignment schemes might be to:

- Reduce defence costs by shortening the overall length of defences to be maintained;
- Increase the efficiency and long term sustainability of flood and coastal defences by recreating river, estuary or coastal habitats and using their flood and storm buffering capacity;
- Provide other environmental benefits through re-creation of natural habitats; or
- Provide replacement habitats in or adjacent to a European designated site to compensate for habitat loss as a result of reclamation or coastal squeeze (Defra 2002, p. 1)

This definition has a broad scope in the opening sentence and is not restrictive in the forms managed realignment may take. However, subsequent definitions are more restrictive and explicitly refer to retreat of the shoreline position (i.e. landward realignment of defences), as illustrated by this quote:

Managed realignment involves the landward movement of a sea defence structure and the promotion of new habitat creation in front of this new line of defence. The land between the old and new defences then forms a new part of the intertidal zone which is more able to respond to coastal processes, and thus reduce the effects of coastal squeeze (French 2004, p. 102).

Through time, the most common understanding of managed realignment became even narrower, including only schemes involving the removal or artificial breaching of flood defences to reinstate tidal flooding into previously defended areas (e.g. French 2006; Wolters et al. 2008; Blackwell et al. 2010; Mossman et al. 2012):

…managed realignment is a technique which is increasingly used to restore intertidal habitat by the removal or breaching of dikes to restore tidal influence…. (Jacobs et al. 2009, p. 368).

This narrow interpretation has led to the popular perception that managed realignment refers exclusively to the landward realignment of the coastal defences, which sometimes are viewed detrimentally as 'giving up land to the sea'.

A different perspective is taken by some practitioners, who consider that managed realignment does not necessarily involve flooding of previously defended land. Under this perspective, it is advocated that managed realignment can involve the creation of habitat by advancing the shoreline seawards. Responding to a survey about perceptions of managed realignment (see Chap. 10), a private consultant from east England wrote: "*Please note managed realignment is not necessarily the same as managed retreat. It is about changing or allowing change within the coastal form… forward or backward*".

The conflicting perceptions about managed realignment and managed retreat result from the inconsistent use of these terms in the literature (see Sect. 2.2). Variations in how managed realignment is interpreted are encouraged by disparities in the definitions presented in Government documents (*cf.* Defra 2002, 2006). The quotation below explicitly refers to realignment of the shoreline forward (i.e. seaward) or backward (i.e. inland), contrasting with other definitions showing a clear focus on concepts of retreat and setback.

...allowing the shoreline to move backwards or forwards, with management to control or limit movement (such as reducing erosion or building new defences on the landward side of the original defences) (Defra 2006, p. 14).

The understanding that managed realignment involves deliberate alterations to existing coastal defences and, therefore requires planning, is common to all interpretations. It is important to recognise the planning element and the creation of multiple benefits to distinguish managed realignment from other initiatives. In some countries (e.g. France, Spain and Portugal), restoration of coastal habitats is more commonly promoted by the natural failure of abandoned defences than through managed realignment. The rationale for planning how and when defences will be altered are indicated here:

> Managed realignment means the deliberate process of altering flood defences to allow flooding of a presently defended area. Managing this process helps to avoid uncertain outcomes and negative impacts and to maximise the potential benefits. Managed realignment may take many forms, dependent on the reasons for undertaking it.... (Leggett et al. 2004, p. 23)

Realising the wider benefits provided by coastlines that are allowed to evolve more naturally and dynamically is central to the concept of managed realignment and sustainable coastal management. Managed realignment is underpinned by the need to manage coastlines in a more sustainable way. Sustainability here might refer to social, economical, environmental or legal aspects (often a complex mix of these elements). As with other soft engineering approaches, the sustainability of managed realignment is based on the adaptive capacity of dynamically evolving coastal habitats and the natural coastal protection (and other ecosystem services) they offer.

2.2 Confusing Terminology

Regional variations and changes in the preferred use of terms through time (see Sect. 2.2) have resulted in the inconsistent use of the terminology in the literature. The terms most commonly used, sometimes as synonyms of managed realignment, include: managed retreat, setback, de-embankment, depoldering, regulated tide exchange (RTE) and controlled reduced tide (CRT). Note that the spelling of some terms also varies, for example:

> ...managed realignment (also known as 'set back', 'managed retreat', or de-poldering in the Netherlands) (French 2006, p. 409).

> Coastal realignment (or managed retreat) is a soft engineering option that aims at re-creating salt marshes and intertidal mudflats by breaching hard coastal defences.... (Badley and Allcorn 2006, p. 102).

> ...managed realignment schemes (also known as de-embankment, de-poldering, set back, wetland mitigation banks, controlled reduced tide (CRT) or flood control area (FCA)) are now in place in many parts of the developed world (Mazik et al. 2010, p. 11).

> The removal of existing flood defences has been variously referred to as managed retreat, managed realignment and habitat creation or restoration, depending on the underlying objectives of the particular scheme. …managed realignment is the form of coastal adaptation that removes a part, or all, of a sea wall in order to allow some additional land area to be subject to tidal action. This may, or may not, require the provision of modified defences, or defences set back on a new line…. (Townend et al. 2010, p. 60).

> Throughout Western Europe and elsewhere, managed realignment schemes (also referred to as depolderisation) are in place…. (Mander et al. 2013, p. 1).

The use of terms is influenced by geographical location and the definition of 'managed realignment' adopted by the author. Although many terms are applied to describe the creation of intertidal areas through artificial restoration of tidal inundation in previously protected land, they are not always recognised as 'managed realignment'. While the term managed realignment is widely used in the UK (and increasingly common in the international literature), it is much less common in other countries.

Outside Europe, managed realignment and its synonyms are not widely used; instead authors may refer to active habitat restoration (e.g. Bowron et al. 2009), tidal hydrology restoration (NOAA 2010) or simply tidal marsh restoration (e.g. Warren et al. 2002; van Proosdij et al. 2010; Brand et al. 2012). However, these terms are not applied exclusively to describe 'managed realignment' projects; it is also used to describe initiatives of habitat restoration purely focused on biodiversity and not related to flood and erosion management.

2.3   Changes in Focus and Terminology Through Time

Pressure from legal requirements to create or enhance intertidal habitat has led to a shift in the focus of managed realignment projects. Initially, sustainable flood risk management seemed to be the primary motivation for managed realignment. However, through time the emphasis has shifted to nature conservation and climate change adaptation. By creating space for coastlines to respond dynamically to changes in environmental conditions, management realignment helps reinstating the natural adaptive capacity of coastal habitats. Therefore, managed realignment is increasingly used as a key mechanism for the restoration of intertidal habitats (e.g. Jacobs et al. 2009) driven by strategic and legislative needs to adapt to sea-level rise, and to compensate for loss and degradation of natural habitats and wildlife.

> Setting back the defences and restoring coastal habitats, known as 'managed realignment', is an important adaptation to rising sea levels. Managed realignment gives coastal habitats space to migrate inland as sea levels rise (Committee on Climate Change 2013, p. 93).

In the UK, the strong emphasis on environmental objectives has created a negative public perception. Public reaction reflects the views that interests and safety of local people and communities are now second to habitat creation, as indicated by the following statements given to the academic survey described in Esteves and Thomas (2014):

National Ocean Council 2013. National Ocean Policy Implementation Plan. http://www.
whitehouse.gov/sites/default/files/national_ocean_policy_implementation_plan.pdf. Accessed
1 Feb 2014.

Parliamentary Office of Science and Technology. (2009). Coastal management, *Postnote*, 342,
p. 5. www.parliament.uk/briefing-papers/POST-PN-342.pdf. Accessed 7 Feb 2014.

Pennings, S. C., & Calloway, R. M. (1992). Salt marsh plant zonation: The relative importance of
competition and physical factors. *Ecology, 73*, 681–690.

Pontee, N. (2013). Defining coastal squeeze: A discussion. *Ocean & Coastal Management, 84*,
204–207.

Shepard, C. C., Crain, C. M., & Beck, M. W. (2011). The protective role of coastal marshes:
A systematic review and meta-analysis. *PLoS One, 6*(11), e27374. doi:10.1371/journal.
pone.0027374.

Smit, B., Burton, I., Klein, R. J. T., & Wandel, J. (2000). An anatomy of adaptation to climate
change and variability. *Climatic Change, 45*, 223–251.

Spalding, M. D., McIvor, A. L., Beck, M. W., Koch, E. W., Möller, I., Reed, D. J., Rubinoff, P.,
Thomas, S., Tolhurst, T. J., Wamsley, T. V., van Wesenbeeck, B. K., Wolanski, E., Woodroffe,
C.D. (2013). Coastal ecosystems: A critical element of risk reduction. *Conservation Letters,
00*, 1–9. doi: 10.1111/conl.12074.

UK National Ecosystem Assessment. (2011). *The UK national ecosystem assessment: Synthesis of
the key findings*. Cambridge: UNEP–WCMC.

van Loon-Steensma, J. M., & Vellinga, P. (2013). Trade-offs between biodiversity and flood protection services of coastal salt marshes. *Current Opinion in Environmental Sustainability, 5*,
320–326.

Yee, A. T. K., Ang, W. F., Teo, S., Liew, S. C., & Tan, H. T. W. (2010). The present extent of man-
grove forests in Singapore. *Nature in Singapore, 3*, 139–145.

References

An, S., Li, H., Guan, B., Zhou, C., Wang, Z., Deng, Z., Zhi, Y., Liu, Y., Xu, C., Fang, S., Jiang, J., & Li, H. (2007). China's natural wetlands: Past problems, current status, and future challenges. *Ambio, 34*, 335–342.

Barbier, E. B., Hacker, S. D., Kennedy, C., Koch, E. W., Stier, A. C., & Silliman, B. R. (2011). The value of estuarine and coastal ecosystem services. *Ecological Monographs, 81*, 169–193.

Dahl, T. E., & Stedman, S. M. (2013). *Status and trends of wetlands in the coastal watersheds of the Conterminous United States 2004–2009* (p. 46). U.S. Department of the Interior, Fish and Wildlife Service and National Oceanic and Atmospheric Administration, National Marine Fisheries Service.

Defra Flood Management Division. (2005). *Coastal squeeze implications for flood management, the requirements of the European Birds and Habitats Directives.* Defra Policy Guidance. https://www.gov.uk/government/uploads/system/uploads/attachment_data/file/181444/coastalsqueeze.pdf. Accessed 3 Feb 2014.

Doody J. P. (2013). Coastal squeeze and managed realignment in southeast England, does it tell us anything about the future? *Ocean & Coastal Management, 79*, 34–41.

Esteves, L. S. (2013). Is managed realignment a sustainable long-term coastal management approach? *Journal of Coastal Research,* Special Issue 65(1), 933–938.

French, P. W. (2004). *Coastal and Estuarine Management.* London: Routledge.

Friess, D. A., Möller, I., Spencer, T., Smith, G. M., Thomson, A. G., & Hill, R. A. (2014). Coastal saltmarsh managed realignment drives rapid breach inlet and external creek evolution, Freiston Shore (UK). *Geomorphology, 208*, 22–33.

Gedan, K. B., Silliman, B. R., & Bertness, M. D. (2009). Centuries of human driven change in salt marsh ecosystems. *Annual Review of Marine Science, 1*, 117–141.

Haines-Young, R. H., & Potschin, M. P. (2010). The links between biodiversity, ecosystem services and human well-being. In D. G. Raffaelli & C. L. J. Frid (Eds.), *Ecosystem ecology: A new synthesis. Cambridge: BES ecological reviews series* (pp. . 110–139). Cambridge: Cambridge University Press.

Hughes, R. G., & Paramor, O. A. L. (2004). On the loss of saltmarshes in south-east England and methods for their restoration. *Journal of Applied Ecology, 41*, 440–448.

IPCC (Intergovernmental Panel on Climate Change). (2013). Summary for policymakers. In T. F. Stocker, D. Qin, G.-K. Plattner, M. Tignor, S. K. Allen, J. Boschung, A. Nauels, Y. Xia, V. Bex, & P. M. Midgley (Eds.), *Climate change 2013: The physical science basis. contribution of working group I to the fifth assessment report of the intergovernmental panel on climate change.* Cambridge: Cambridge University Press.

Liquete, C., Piroddi, C., Drakou, E. G., Gurney, L., Katsanevakis, S., Charef, A., & Egoh, B. (2013). Current status and future prospects for the assessment of marine and coastal ecosystem services: A systematic review. *PLoS One, 8*(7), e67737. doi:10.1371/journal.pone.0067737.

MacDonald, J. M., & Harris, N. (Ed.) (2013). Proceedings of the 8th Annual Elwha Nearshore Consortium Workshop (27 February 2013), Port Angeles, Washington. http://www.clallam.net/ccmrc/documents/ENC_2013_Proceedings.pdf. Accessed 10 Feb 2014.

Maltby, E. (2006). Wetland conservation and management: Questions for science and society in applying the ecosystem approach. In: R. Bobbink, B. Beltman, J. T. A. Verhoeven, & D. F. Whigham (Eds.), *Wetlands: Functioning, biodiversity conservation, and restoration.* Ecological Studies, 191(2), 93–116.

Martínez, M. L., Intralawan, A., Vázquez, G., Pérez-Maqueo, O., Sutton, P., Landgrave, R. (2007). The coasts of our world: Ecological, economic and social importance. *Ecological Economics, 63*, 254–272.

Millennium Ecosystem Assessment (2005). Ecosystems and human well-being: Wetlands and water—Synthesis. Washington: Island Press.

Möller, I. and Spencer, T. (2002) Wave dissipation over macro-tidal saltmarshes: Effects of marsh edge typology and vegetation change. *Journal of Coastal Research,* Special Issue 36, 506–521.

A useful tool has become a plaything for environmentalists (stakeholder from east England, on 1 Aug 2013).

this realignment for the birds, is barmy… (member of the public from east England, on 1 Aug 2013).

Madness!! when we have starving people in the world and we want to flood productive land! (Farmer from east England, on 2 Aug 2013).

…why do various so called environmentalists think that they can 'play God' in deciding that land that is home to a huge variety of land animals, plants and crops, insects and birds is all of a sudden flooded for some wading birds! (Stakeholder from east England, on 2 Aug 2013).

…most sites today, however conceived, will ultimately look like nature conservation sites for birds. This image puts many farmers off, and even causes antagonism (Consultant from south England, on 7 Aug 2013).

…too much emphasis is on habitat creation only approaches and this constrains the opportunities for more societal benefits and sometimes stops projects progressing as it is seen as people v birds. …The approach needs to be remarketed and 'sold' more effectively … so that landowners and the coastal communities want it rather than assuming …it's a bad thing/giving up to the sea (government practitioner from east England, on 6 Aug 2013).

In an attempt to disentangle negative connotations associated with managed realignment, the terminology used by government and practitioners in the UK has evolved through time. Managed retreat and setback were commonly used in earlier documents, but have gradually fallen in disuse for being interpreted as 'giving up land to the sea'. This interpretation was perceived as a government failure or slackness in providing protection against flooding and erosion, as evidenced in the literature:

…managed realignment... may also be referred to as managed retreat or set back. Increasingly, the term managed retreat is going out of favour as it suggests a negativity in coastal management, 'retreating in the face of the enemy (sea)' which many coastal managers and coastal residents find unacceptable (French 2001, p. 271).

The restoration of tidal wetlands in the United Kingdom was initially referred to as "setback," a pejorative phrase that was rapidly changed, first to "managed retreat" but that is now officially termed "managed realignment" (Pethick 2002, p. 431).

In many locations worldwide, hard engineering has been the type of coastal defence most people expect. Therefore, managed realignment represents a shift from the *status quo*. How people react to change in general depends on the potential threats it might bring to the individual and collective way of life, how and from whom they learn about the change and their level and type of engagement (e.g. Lorenzone et al. 2007; Pidgeon et al. 2008; O'Neill and Nicholson-Cole 2009; Hobson and Niemeyer 2011).

It is much harder for people to accept change if their initial perception is associated with a negative impact or connotation. In the case of government policies, for example, it is important that a clear and consistent message is used to explain the changes they might bring to individuals and society. As importantly, the terminology associated with new policies must be carefully considered to reduce the chances of undesirable interpretations.

The lack of a clear understanding about the definition of terms, policy drivers and outcomes can lead to avoidable misunderstandings among practitioners, researchers and, particularly, the public. 'Negative' terminology is thought to have caused such an impact on the wider perception of managed realignment in the UK that once again the government is assessing whether a new term should be used (see Chap. 8). Therefore, it is possible that managed realignment will be substituted by a new term in future documents.

2.4   Clarifying the Terminology

De-embankment and depolderisation[1] are used, especially in northern Europe, to describe total or partial removal (breaching) of flood defences to create intertidal areas in previously embanked land (e.g. Rupp-Armstrong and Nicholls 2007; Mander et al. 2013). Note that in France[2], term *dépoldériser* is similarly used (e.g. Goeldner-Gianella 2007). De-embankment sometimes refers also to accidental or unplanned breaching of flood defences (e.g. during storms) and adjectives such as 'deliberate' de-embankment might be used to differentiate (e.g. Wolters et al. 2005).

Setback is used as a synonym of managed realignment only by few authors (e.g. French 2006; Ducrotoy and Elliott 2006; Mazik et al. 2010). The term is said to have fallen into disuse owing to its negative connotations (e.g. French 2001; Elliott and Cutts 2004). In the context of managed realignment, setback means realignment of the defence line to an inland position. However, in the literature the term setback most commonly refers to construction control zoning to keep buildings and people away from hazard zones (e.g. Sanò et al. 2011; Portman et al. 2012; Ramsay et al. 2012; Abbott 2013; Mycoo 2013; Reisinger et al. 2014).

Setback lines or areas indicate the minimum distance from the shoreline new development must be built (Fig. 2.1), which can be defined based on fixed distance from a selected shoreline proxy (e.g. high water line, cliff edge, dune toe), historic erosion rates or the extent of extreme flooding. For example, the Article 8-2 of the Mediterranean Integrated Coastal Zone Management Protocol[3] (signed in 2008) establishes that signatory countries should create a setback zone of at least 100 m in width from the highest winter waterline where no constructions are allowed. In Hawaii, the shoreline setback is determined based on the likely shoreline retreat to occur during the life-time of the building (see Chap. 7). Shoreline setback is often implemented as a mechanism to support managed retreat, as illustrated in Chap. 7.

---

[1] Low-lying areas, reclaimed from the sea and protected by embankments are called *polders*, a word of Dutch origin commonly used in northern Europe.

[2] See, for example, the debate and documents about the project at Hable d' Ault, at the coast of Picardy at: http://www.baiedesomme.org/actu/hable-dault-depolderiser-or-not-depolderiser-424.html.

[3] The text of the protocol and other associated documents are available from: http://www.pap-thecoastcentre.org/itl_public.php?public_id=365&lang=en.

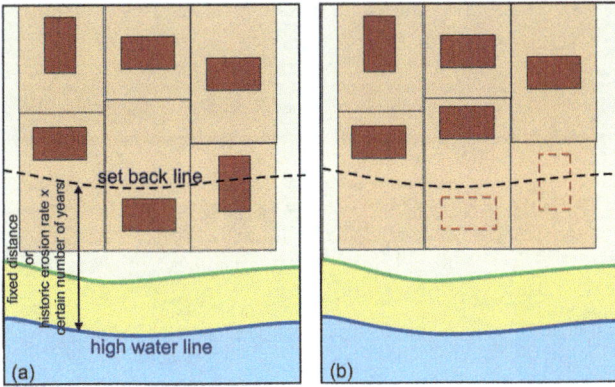

Fig. 2.1 (**a**) A setback line defines an area where constructions are not allowed with the aim to safeguard people and property from erosion and/or flooding. (**b**) If buildings placed seaward of the setback line need to be re-developed (e.g. due to impact of storms), this must only be permitted inland of the setback line, within the same land lot, if space is available, or elsewhere

Managed retreat and managed realignment are used interchangeably in the UK (e.g. Emmerson et al. 1997; Chang et al. 1998; Macleod et al. 1999; French 2001; Pethick 2002; Cooper 2003), more commonly in the literature pre-dating the publication of the policy *Making Space for Water* (Defra 2005). Elsewhere, such as in Spain (Roca et al. 2008), the USA (Siders 2013), Australia (Alexander et al. 2012) and New Zealand (Turbott and Stuart 2007; Reisinger et al. 2014), managed retreat most often refers to the relocation of property at risk and/or allowing the shoreline to move more dynamically (e.g. Jackson and Nordstrom 2013).

> Managed retreat—the relocation of homes and infrastructure under threat from coastal flooding—is one of the few policy options available for coastal communities facing long-term risks from accelerated sea level rise. (Alexander et al. 2012, p. 409)

> Retreat, which works with natural dynamics and leaves more space for water and sediment. Infrastructures are removed and land uses can be abandoned. (Roca et al. 2008, p. 406)

Regulated tidal exchange (RTE) is a term used to describe projects of habitat restoration where a controlled tidal flow (extent and duration) is reinstated into embanked areas through culverts and sluices. This approach is widely used worldwide but the term RTE is most commonly used in Europe. In France, for example, RTE was implemented in 1999 to reinstate tidal flows into the 132 ha of the Polder de Sébastopol, Île de Noirmoutier, Vendée (Fig. 2.2). The area was purchased by the local government in 1986 and a project of nature restoration was started in 1996 to re-create intertidal habitats for the protection of birds. In 2008 the area was designated as Regional Natural Reserve[4] of the Pays de la Loire region.

---

[4] Information about the Regional Nature Reserve is available at: http://www.reserves-naturelles.org/polder-de-sebastopol.

Fig. 2.2 The Polder de Sébastopol (Vendée, France), reclaimed from the sea in 1856, had tidal flows restored into the embanked area through regulated tidal exchange implemented in 1999 (photo: Jacques Oudin, courtesy of Communauté de Communes de île de Noirmoutier)

As RTE sites will still be protected from wave impact by the flood defence, and tidal flow is controlled, sediment deposition is enhanced thereby maximising the chances for saltmarsh development, especially at low-lying locations (Nottage and Robertson 2005). Culverts can be designed to reduce the flooding height at sites with a low elevation, but the spring-neap variation in water level tends to be greatly reduced, limiting the restoration of the full spectrum of intertidal gradient (Beauchard et al. 2011). In some instances, RTE schemes are excluded from analyses of managed realignment projects (e.g. Rupp-Armstrong and Nicholls 2007; Esteves 2013), while in other cases, although differences are recognised, RTE is considered a form of managed realignment:

> regulated tidal exchange (RTE)... differs from managed realignment in that tidal flooding enters the site through tidal gates or sluices, leaving the sea wall intact (hereafter RTE sites are included as [managed realignment]) (Mossman et al. 2012, p. 1447).

Flood control areas (FCAs) are, in essence, spaces used to 'store' overflow waters with the objective of reducing the risk of flooding elsewhere. Floodplains and coastal plains naturally have the function of flood control; however, development has greatly reduced this natural capacity. Areas that once contributed to flood risk reduction have since lost their capacity to do so and now require protection from flooding themselves. FCAs are designed to create a high storage-capacity to contain floodwater in relatively small areas. Often, FCAs are enclosed by flood defences, which usually have lower embankments or dykes in the section fronting the river

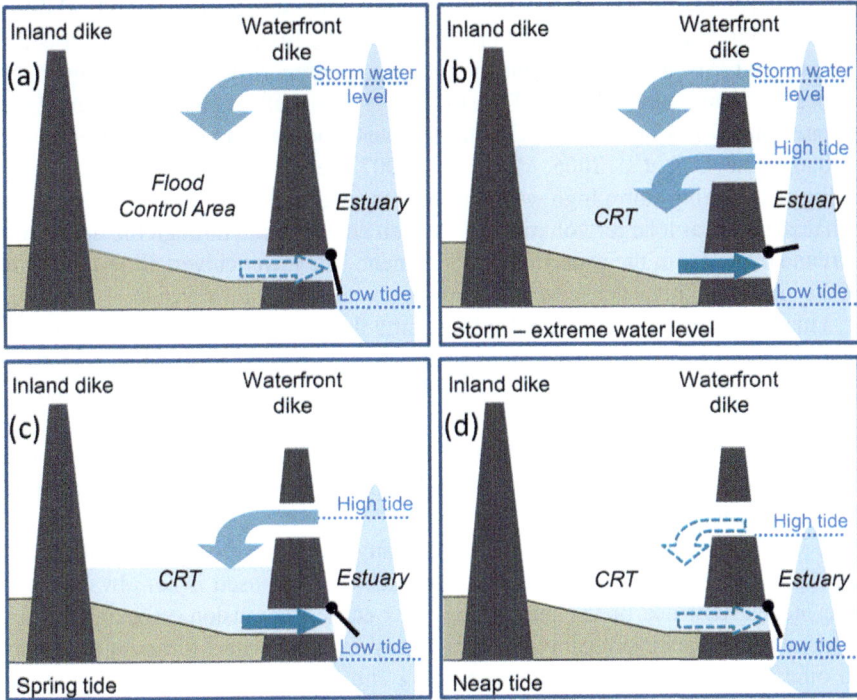

Fig. 2.3 (**a**) Diagram representing the water flow exchange within flood control areas and control reduced tide schemes, (**b**) during storm water levels, (**c**) spring tides and (**d**) neap tides. Depending on the water level, the water enters the site through the upper culvert and/or over the waterfront dike. The water leaves the site through the lower culvert (controlled by an outlet valve) when water level outside is below the water level within the site (Modified from: (**a**) Cox et al. (2006) and (**b–d**) Beauchard et al. (2011) and Beauchard (2012))

or coast and higher dykes elsewhere (Cox et al. 2006). During high water events (e.g. river overflow, storm surges), water overflowing from the lower waterfront defences is contained by the higher inland defences (Fig. 2.3a) thereby reducing the risk of flooding further inland and/or upstream (Meire et al. 2005).

Although managed realignment creates space and enhances the capacity for natural flood control, FCAs *per se* do not necessarily represent a mechanism of managed realignment and the term is rarely used in this context. The frequency of flooding in FCAs may be only once or twice a year depending on extreme weather events and therefore the chances for development of intertidal habitat is limited (Cox et al. 2006). Exceptions occur when FCAs are managed to enhance opportunities for the delivery of other benefits through the development of intertidal habitats as currently practiced in Belgium.

Controlled reduced tide (CRT) is an approach similar to RTE, more widely applied in Belgium to describe the use of carefully designed sluices to control tidal flows within FCAs with the objective of promoting intertidal habitat development (Cox et al. 2006; Maris et al. 2007; Jacobs et al. 2009; Vandenbruwaene et al. 2011;

Beauchard et al. 2011; Beauchard et al. 2013). In CRT schemes, as the inland flood defences will still limit the inland progression of the highest water levels, it is the landward edge of intertidal habitats, not the flood defence, line that is realigned.

In a CRT, the water enters the FCA through an inlet culvert built high in the flood defence and flows out of the FCA through an outlet culvert placed lower in the embankment (Cox et al. 2006; Maris et al. 2007; Beauchard et al. 2011), as shown in Fig. 2.1. During storm high water conditions (Fig. 2.3b), the area can work as an enhanced FCA as a larger volume of water can enter the site through the upper culvert and overflow of the waterfront embankment. The lower culvert allows draining of the site when the water level outside is below the water level inside.

This design creates an internal tidal regime that maintains a spring-neap variation (Fig. 2.3c), which can be modulated based on habitat restoration objectives (Beauchard et al. 2011). Inundation depths tend to be reduced and durations increased in comparison with adjacent tidal marsh areas (Cox et al. 2006; Maris et al. 2007). This altered tidal regime allows marshes to develop on land with elevations much lower than would be possible under a normal tidal regime (Beauchard et al. 2011; Vandenbruwaene et al. 2011).

It is evident now that in the literature, many terms have been associated with managed realignment and the context in which they are used is not always clear. The inconsistent use of the terminology has caused confusion, especially when terms are used interchangeably on some occasions, and reflect different meanings on others. To facilitate wider understanding and stimulate consistency in the use of terminology, it is suggested here that the term managed realignment is applied more broadly to indicate a group of approaches that can be implemented to create opportunities for coasts to evolve more dynamically and be managed more sustainably. A revised definition for managed realignment is suggested in the next section.

2.5   Proposing a New Definition

Taking into consideration the various definitions and aspects influencing the general understanding of managed realignment, a less prescriptive but more widely applicable definition may be: managed realignment is a soft engineering approach aiming to promote (socio-economic, environmental and legal) sustainability of coastal erosion and flood risk management by creating opportunities for the realisation of the wider benefits provided by the natural adaptive capacity of coastlines that are allowed to respond more dynamically to environmental change. Therefore, 'managed' refers to take purposefully actions, to plan, implement and monitor projects; and 'realignment' refers to the position of the shoreline and/or the line of defences.

Within this broader context, managed realignment becomes a general term that can be used to describe collectively the many mechanisms implemented to allow coastlines to evolve more flexibly. In this regard, creating space for enhancing the adaptive capacity of coastlines can be achieved by either landward or seaward shoreline realignment. Chapter 3 describes five categories of methods that can be used to implement managed realignment: removal of flood defences, breach of defences, realignment of defences, controlled tidal flow and managed retreat.

The distinction then between managed realignment and other soft engineering approaches is that decisions are made based on an integrated long-term planning for the delivery of multiple functions (and the associated tangible benefits) is intrinsic to the process, which should take into consideration the uncertainties related to the variability of naturally evolving coasts. Adopting the definition of managed realignment suggested here will help clarify the terminology, which is currently used inconsistently, leaving margin to conflicting and varied interpretations.

References

Abbott, T. (2013). Shifting shorelines and political winds—The complexities of implementing the simple idea of shoreline setbacks for oceanfront developments in Maui, Hawaii. *Ocean & Coastal Management, 73*, 13–21.

Alexander, K. S., Ryan, A., & Measham, T. G. (2012). Managed retreat of coastal communities: Understanding responses to projected sea level rise. *Journal of Environmental Planning and Management, 55*(4), 409–433.

Badley, J., & Allcorn, R. I. (2006). Changes in bird use following the managed realignment at Freiston Shore RSPB Reserve, Lincolnshire, England. *Conservation Evidence, 3*, 102–105.

Beauchard, O. (2012). Tidal freshwater habitat restoration through controlled reduced tide system: A multi-level assessment. Ph.D. thesis, University of Antwerp, University Press Antwerp.

Beauchard, O., Jacobs, S., Cox, T. J. S., Maris, T., Vrebos, D., Van Braeckel, A., & Meire, P. (2011). A new technique for tidal habitat restoration: Evaluation of its hydrological potentials. *Ecological Engineering, 37*, 1849–1858.

Beauchard, O., Jacobs, S., Ysebaert, T., & Meire, P. (2013). Sediment macroinvertebrate community functioning in impacted and newly-created tidal freshwater habitats. *Estuarine, Coastal and Shelf Science, 120*, 21–32.

Blackwell, M. S. A., Yamulki, S., & Bol, R. (2010). Nitrous oxide production and denitrification rates in estuarine intertidal saltmarsh and managed realignment zones. *Estuarine, Coastal and Shelf Science, 87*(4), 591–600.

Bowron, T., Neatt, N., van Proosdij, D., Lundholm, J., & Graham, J. (2009). Macro-tidal salt marsh ecosystem response to culvert expansion. *Restoration Ecology, 19*(3), 307–322.

Brand, L. A., Smith, L. M., Takekawa, J. Y., Athearn, N. D., Taylor, K., Shellenbarger, G. G., Schoellhamer, D. H., & Spenst, R. (2012). Trajectory of early tidal marsh restoration: Elevation, sedimentation and colonization of breached salt ponds in the northern San Francisco Bay. *Ecological Engineering, 42*, 19–29.

Chang, Y. H., Scrimshaw, M. D., Emmerson, R. H. C., & Lester, J. N. (1998). Geostatistical analysis of sampling uncertainty at the Tollesbury Managed Retreat site in Blackwater Estuary, Essex, UK: Kriging and cokriging approach to minimise sampling density. *Science of the Total Environment, 221*, 43–57.

Committee on Climate Change. (2013). Managing the land in a changing climate. Chapter 5: Regulating services—Coastal habitats (pp. 92–107). http://www.theccc.org.uk/publication/managing-the-land-in-a-changing-climate/. Accessed 1 Dec 2013.

Cooper, N. J. (2003). The use of 'managed retreat' in coastal engineering. *Proceedings of the institution of civil engineering, engineering sustainability, 156*(ES2), 101–110.

Cox, T., Maris, T., De Vleeschauwer, P., De Mulder, T., Soetaert, K., & Meire, P. (2006). Flood control areas as an opportunity to restore estuarine habitat. *Ecological Engineering, 28*, 55–63.

Defra (Department for Environment, Food and Rural Affairs). (2002). Managed realignment review. Policy research project FD 2008, Flood and Coastal Defence R & D Programme. http://randd.defra.gov.uk/Document.aspx?Document=FD2008_537_TRP.pdf. Accessed 10 Jan 2014.

Defra (Department for Environment, Food and Rural Affairs). (2005). *Making space for water— Taking forward a new Government strategy for flood and coastal erosion risk management in England*. First Government response to the autumn 2004 consultation exercise. London.

Defra (Department for Environment, Food and Rural Affairs). (March, 2006). Shoreline management plan guidance. Volume 1: Aims and requirements. https://www.gov.uk/government/publications/shoreline-management-plans-guidance. Accessed 29th Dec 2013.

Ducrotoy, J. P., & Elliott, M. (2006). Recent developments in estuarine ecology and management. *Marine Pollution Bulletin, 53,* 1–4.

Elliott, M., & Cutts, N. D. (2004). Marine habitats: Loss and gain, mitigation and compensation. *Marine Pollution Bulletin, 49,* 671–674.

Emmerson, R. H. C., Manatunge, J. M. A., Macleod, C. L., & Lester, J. N. (1997). Tidal exchanges between orplands managed retreat site and the blackwater estuary, Essex. *Water and Environment Journal, 11,* 363–372.

Esteves, L. S. (2013). Is managed realignment a sustainable long-term coastal management approach? *Journal of Coastal Research, Special Issue, 65*(1), 933–938.

Esteves, L. S., & Thomas, K. (2014). Managed realignment in practice in the UK: Results from two independent surveys. *Journal of Coastal Research, Special Issue, 70,* 407–413.

French, P. W. (2001). *Coastal defences: Processes, problems and solutions.* London: Routledge.

French, P. W. (2004). *Coastal and estuarine management.* London: Routledge.

French, P. W. (2006). Managed realignment—The developing story of a comparatively new approach to soft engineering. *Estuarine, Coastal and Shelf Science, 67*(3), 409–423.

Goeldner-Gianella, L. (2007). Dépoldériser en Europe occidentale. *Annales de géographie, 656,* 339–360.

Hobson, K., & Niemeyer, S. (2011). Public responses to climate change: The role of eliberation in building capacity for adaptive action. *Global Environmental Change, 21*(3), 957–971.

Jackson, N. L., & Nordstrom, K. F. (2013). Removing shore protection structures to facilitate migration of landforms and habitats on the bayside of a barrier spit. *Geomorphology, 199,* 179–191.

Jacobs, S., Beauchard, O., Struyf, E., Cox, T., Maris, T., & Meire, P. (2009). Restoration of tidal freshwater vegetation using controlled reduced tide (CRT) along the Schelde Estuary (Belgium). *Estuarine, Coastal and Shelf Science, 85*(3), 368–376.

Leggett, D. J., Cooper, N., & Harvey, R. (2004). *Coastal and estuarine managed realignment— Design.* CIRIA C628D. London: CIRIA, 215 p (ISBN: 978-0-86017-628-2).

Lorenzone, I., Nicholson-Cole, S., & Whitmarsh, L. (2007). Barriers perceived to engaging with climate change among the UK public and their policy implications. *Global Environmental Change, 17,* 445–459.

Macleod, C. L., Scrimshaw, M. D., Emmerson, R. H. C., Chang, Y. H., & Lester, J. N. (1999). Geochemical changes in metal and nutrient loading at orplands farm managed retreat site, Essex, UK (April 1995–1997). *Marine Pollution Bulletin, 38,* 1115–1125.

Mander, L., Marie-Orleach, L., & Elliott, M. (2013). The value of wader foraging behaviour study to assess the success of restored intertidal areas. *Estuarine, Coastal and Shelf Science, 131,* 1–5.

Maris, T., Cox, T., Temmerman, S., De Vleeschauwer, P., Van Damme, S., De Mulder, T., Van den Bergh, E., & Meire, P. (2007). Tuning the tide: Creating ecological conditions for tidal marsh development in a controlled inundation area. *Hydrobiologia, 588,* 31–43.

Mazik, K., Musk, W., Dawes, O., Solyanko, K., Brown, S., Mander, L., & Elliott, M. (2010). Managed realignment as compensation for the loss of intertidal mudflat: A short term solution to a long term problem? *Estuarine, Coastal and Shelf Science, 90*(1), 11–20.

Meire, P., Ysebaert, T., Van Damme, S., Van den Bergh, E., Maris, T., & Struyf, E. (2005). The Scheldt estuary: A description of a changing ecosystem. *Hydrobiologia, 540,* 1–11.

Mossman, H. L., Davy, A. J., & Grant, A. (2012). Does managed coastal realignment create saltmarshes with 'equivalent biological characteristics' to natural reference sites? *Journal of Applied Ecology, 49,* 1446–1456.

Mycoo, M. (2014). Sustainable tourism, climate change and sea level rise adaptation policies in Barbados. *Natural Resources Forum, 36,* 47–57.

NOAA (Restoration Center and Coastal Services Center). (2010). *Returning the tide—A tidal* hydrology restoration guidance manual for the Southeastern U.S. NOAA. MD: Silver Spring. http://www.habitat.noaa.gov/partners/toolkits/tidal_hydro.html. Accessed 20 Feb 2014.

Nottage, A., & Robertson, P. (2005). *The saltmarsh creation handbook: A project manager's guide to the creation of saltmarsh and intertidal mudflat*. London: The Royal Society for the Protection of Birds and CIWEM.

O'Neill, S., & Nicholson-Cole, S. (2009). "Fear Won't Do It": Promoting positive engagement with climate change through visual and iconic representations. *Science Communication, 30*, 355–379.

Pethick, J. (2002). Estuarine and tidal wetland restoration in the United Kingdom: Policy versus practice. *Restoration Ecology, 10*(3), 431–437.

Pidgeon, N. F., Simmons, P., Sarre, S., Henwood, K. L., & Smith, N. (2008). The ethics of socio-cultural risk research. *Health, Risk & Society, 10*(4), 321–329.

Portman, M. E., Esteves, L. S., Le, X. Q., & Khan, A. Z. (2012). Improving integration for integrated coastal zone management: An eight country study. *Science of the Total Environment, 439*, 194–201.

Ramsay, D. L., Gibberd, B., Dahm, J., & Bell, R. G. (2012). *Defining coastal hazard zones and setback lines. A guide to good practice*. Hamilton: National Institute of Water & Atmospheric Research Ltd.

Reisinger, A., Lawrence, J., Hart, G., & Chapman, R. (2014). From coping to resilience: The role of managed retreat in highly developed coastal regions of New Zealand. In B. Glavovic, R. Kaye, M. Kelly, & A. Travers (Eds.), *Climate change and the coast: Building resilient communities*. London: CRC Press.

Roca, E., Gamboa, G., & Tàbara, J. D. (2008). Assessing the multidimensionality of coastal erosion risks: Public participation and multicriteria analysis in a Mediterranean Coastal System. *Risk Analysis, 28*(2), 399–412.

Rupp-Armstrong, S., & Nicholls, R.J. (2007). Coastal and estuarine retreat: A comparison of the application of managed realignment in England and Germany. *Journal of Coastal Research, 23*(6), 1418–1430.

Sanò, M., Jimenez, J. A., Medina, R., Stanica, A., Sanchez-Arcilla, A., & Trumbic, I. (2011). The role of coastal setbacks in the context of coastal erosion and climate change. *Ocean & Coastal Management, 54*, 943–950.

Siders, A. (2013). Managed coastal retreat: A legal handbook on shifting development away from vulnerably areas. Centre for Climate Change Law, Columbia Public Law Research Paper No. 14–365, p. 140.

Townend, I., Scott, C., & Dixon, M. (2010). Managed realignment: A coastal flood management strategy. In G. Pender & H. Faulkner (Eds.), *Flood science risk and management* (pp. 60–86). Oxford: Wiley.

Turbott, C., & Stuart, A. (2007). Managed retreat from coastal hazards: Options for implementation. Environment Waikato technical report 2006/48, p. 89. http://www.waikatoregion.govt.nz/PageFiles/5405/tr06-48.pdf. Accessed 20 Jan 2014.

Vandenbruwaene, W., Maris, T., Cox, T. J. S., Cahoon, D. R., Meire, P., & Temmerman, S. (2011). Sedimentation and response to sea-level rise of a restored marsh with reduced tidal exchange: Comparison with a natural tidal marsh. *Geomorphology, 130*, 115–126.

van Proosdij, D., Lundholm, J., Neatt, N., Bowron, T., & Graham J. (2010). Ecological re-engineering of a freshwater impoundment for salt marsh restoration in a hypertidal system. *Ecological Engineering, 36*(10), 1314–1332.

Warren, R. S., Fell, P. E., Rozsa, R., Brawley, A. H., Orsted, A. C., Olson, E. T., Swamy, V., & Niering, A. (2002). Salt marsh restoration in Connecticut: 20 years of science and management. *Restoration Ecology, 10*(3), 497–513.

Wolters, M., Garbutt, A., & Bakker, J. P. (2005). Salt-marsh restoration: Evaluating the success of de-embankments in north-west Europe. *Biological Conservation, 123*(2), 249–268.

Wolters, M., Garbutt, A., Bekker, R. M., Bakker, J. P., & Carey, P. D. (2008). Restoration of salt-marsh vegetation in relation to site suitability, species pool and dispersal traits. *Journal of Applied Ecology, 45*(3), 904–912.

Chapter 3
Methods of Implementation

Luciana S. Esteves

Abstract This chapter describes the different methods most commonly used in the implementation of managed realignment. Following the wider definition of managed realignment proposed in Chap. 2, the implementation of managed realignment is more widespread than one might initially anticipate. Many approaches described in the literature reflect different ways in which managed realignment can be implemented. These approaches are grouped into five methods of implementation: removal of defences; breach of defences; realignment of defences; controlled tidal restoration (which includes regulated tidal exchange and controlled reduced tide) and managed retreat. This chapter provides examples of implementation worldwide and describes and contrasts each one of these methods.

3.1 Introduction

The objective of this chapter is to describe the different methods most commonly used in the implementation of managed realignment. In Chap. 2 a broad definition of managed realignment is proposed. Following this definition, many approaches described in the literature reflect different ways in which managed realignment can be implemented. Here these approaches are grouped into five methods of implementation: removal of defences; breach of defences; realignment of defences; controlled tidal restoration and managed retreat (Table 3.1). The wider adoption of the definition and terminology proposed here would reduce the many conflicting uses and facilitate the wider understanding of the managed realignment concept and the differences between the methods of implementation.

Table 3.1 identifies the primary and secondary elements characterising each method of implementation. The primary characteristic is the main action taken in the implementation of the project, which generally is reflected in the name of the implementation method. The secondary characteristic depends mostly on the site's pre-existing conditions (e.g. topography, land use, presence of a secondary line of

L. S. Esteves (✉)
Faculty of Science and Technology, Bournemouth University, Talbot Campus,
Poole, Dorset, BH12 5BB, UK
e-mail: lesteves@bournemouth.ac.uk

L. S. Esteves, *Managed Realignment: A Viable Long-Term Coastal Management Strategy?*,   33
SpringerBriefs in Environmental Science, DOI 10.1007/978-94-017-9029-1_3,
© Springer Science+Business Media Dordrecht 2014

Table 3.1 Primary and secondary characteristics of the five managed realignment methods of implementation (black = primary; grey = secondary; white = not applicable)

	Managed realignment methods of implementation					
	Removal of defences	Breach of defences	Realignment of defences	Controlled tidal restoration		Managed retreat
				RTE	CRT	
Extended sections of coastal defences are removed	primary	n/a	secondary			
Coastal defence is artificially breached						
Defence is allowed to breach naturally						
Project involves new line of defence or upgrading existing defences						
Sluices and culverts restore a controlled tidal flow						
Project involves flood control areas						
Planned removal of people and assets at risk						
Primary and secondary (short-term) outcomes						
Creation of habitat						
Improved flood risk management		*		*		
Other ecosystem services						
Climate change adaptation						
Potential for application in						
Urban areas	low to moderate					
Areas of low occupation		high				
Rural areas						

* Improved flood risk depends on the habitat that will be created and therefore it should be considered either a secondary outcome or a long-term primary objective.

defence etc.). Table 3.1 indicates whether each method is more likely to have habitat creation or erosion/flood risk management as a primary outcome and their potential suitability to urban and rural areas (low occupation might refer to low population density and/or fewer properties). It is important to note that many managed realignment projects have multiple objectives and the priorities may vary between projects. Here the identification of primary or secondary outcome refers to the most immediate outcomes. By defining the short- and long-term objectives, it is possible then to select the most adequate method of implementation.

The identification of the most immediate outcome of each method (as shown in Table 3.1) is based on two simple assumptions. In the methods involving creation of space for a more naturally evolving coastline without the construction of new or upgraded defences, reduction of flood risk will change depending on how the coast (and the habitat created) will evolve; therefore, flood risk management is considered a secondary outcome. Flood risk management is a primary outcome of methods that create opportunity for a more naturally evolving coast through controlled conditions and/or confined within defences. Flood risk management is also a primary outcome of managed retreat, as it aims to reduce the number of people, property and infrastructure at risk.

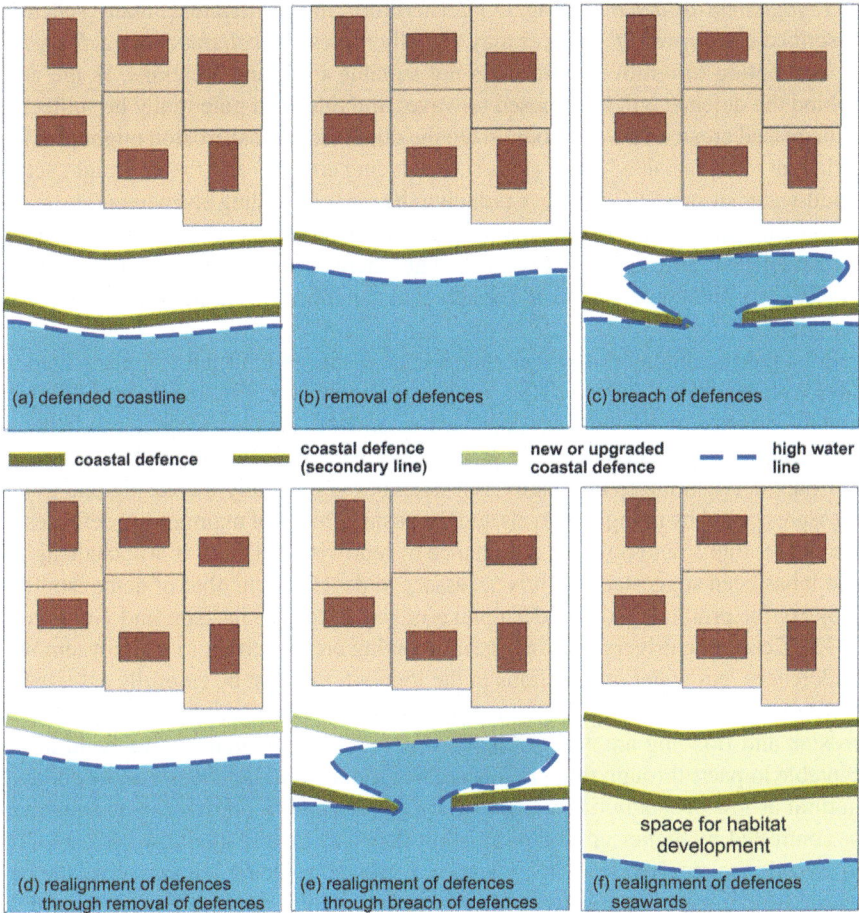

Fig. 3.1 Often when the location is backed by higher topography or (**a**) a secondary line of the defence, (**b**) managed realignment is implemented through removal of defences or (**c**) breach of defences. (**d**) Realignment of defences, most commonly involves building or upgrading a defence line further inland, and the removal or (**e**) breach of an existing line of defence. (**f**) It is proposed here that seaward realignment of defences may also be considered a method of managed realignment if aiming to improve the sustainability of flood and erosion risk management by creating space for the development of habitat

3.2 Removal of Defences

Entire sections of defences are removed (Fig. 3.1) allowing the coastline to respond more freely to waves and tides (i.e. resulting in the realignment of the shoreline position). To be considered a managed realignment implemented through *removal of defences*, building new or upgraded defences should not be included as part of the project design. In cases where the removal of defences is planned in combination with the construction of new or upgraded defences further inland are described here as *realignment of defences* (Fig. 3.1). Therefore, in the literature, many schemes described as *removal of defences* may actually represent *realignment of defences*.

The option to remove defences is not suitable for many locations, as the area behind the defence will be exposed to waves and tides and potentially be subjected to increased erosion and/or flooding. On the other hand, local erosion might provide sediment to replenish adjacent areas (i.e. reducing erosion) and the site might act as flood-water storage (FCA effect), reducing the risk of flooding elsewhere. The managed realignment site will then act as a sacrificial area to create benefits elsewhere. Depending on the characteristics of the site, there is also the potential to enhance biodiversity and create intertidal habitats that can offer a wider range of ecosystem services.

Managed realignment through removal of defences is usually implemented at sites with particular characteristics, usually involving a willing landowner. The National Trust has promoted such approach as a more sustainable option to manage its coastlines, as illustrated by the Brownsea Island project (Dorset, UK), detailed in Chap. 6. Attention to the removal of defences is gradually increasing due to the pressures posed by rising sea levels and the sustainability of maintaining defences in the future. Public and political acceptance of removing defences is still challenging and it has been suggested that it is necessary to develop a number of demonstration projects for promoting the wider uptake of this strategy (Jackson and Nordstrom 2013). Economic drivers and long-term planning are important to support removal of defences, but wider implications in the longer term must be carefully assessed.

Removal of defences is likely to be implemented at locations where: (a) local erosion and flooding are acceptable; and (b) risks can be managed to avoid unacceptable impacts through time. In this regards, removal of defences is more suitable in rural or low occupation areas and where erosion and flood risk can be contained or controlled by higher ground or existing defences further inland, which are able to provide the required level of protection without upgrading. To maximise the potential for realising benefits and minimising detrimental impacts, the planning processes must take careful consideration about: the intended objectives; expectations about potential gains and losses and the associated time-frames in which these are likely to occur; the characteristics of the site, and how they might affect the evolution of coastal change in the short- and long-term. For example, this method of managed realignment might not be the most suitable for the creation of saltmarshes as the increased exposure to tidal currents and waves might prevent the development of such habitats (Nottage and Robertson 2005).

In Europe, the earliest deliberate removal of defences to restore estuarine habitat was implemented at l'Aber de Crozon[1] (Brittany, France). The construction of a dyke (*la digue* Rozan) in 1860 reclaimed part of the estuary to create agricultural land and another dyke (*la digue* Richet) was built in 1958 reclaiming a further 40.5 ha to create a harbour. Changes in economic and environmental policies led

---

[1] As reported by *Le Conservatoire du Littoral* at: http://www.conservatoire-du-littoral.fr/siteLittoral/57/28-l-aber-29-_finistere.htm.

to the removal of *la digue* Richet in 1981 and alterations of *la digue* Rozan by the *Conservatoire du littoral* with the objective of restoring 87 ha of intertidal habitat.[2]

3.3   Breach of Defences

This method is similar to removal of defences, but here only selected sections of the existing defences are removed to allow tides and waves into previously protected land (Fig. 3.1). Therefore, the considerations about the suitability of sites for the removal of defences are also applicable to managed realignment through breaching of defences. However, this alternative maintains certain degree of shelter behind the remaining defences reducing the effects of waves and tides. The sheltering effect might be desired to promote sedimentation and favour the development of saltmarshes, for example. In the lack of maintenance, the remaining of the defence will eventually degrade and the effects of increased exposure to the newly created habitat must be taken into consideration in the longer term (Nottage and Robertson 2005). However, if sheltering is essential, it is possible to keep regular maintenance in selected sections of the remaining defences.

Managed realignment schemes may involve one or multiple breaches. Often these schemes also involve building a new line of defences further inland and, therefore, these cases would be more correctly categorised as *realignment of defences*. Therefore, in the literature, many schemes described as breaching of defences may actually represent realignment of defences. Expert planning of the design and location of breaches is the key element controlling the tidal flow and level of exposure within the realignment site, which in turn influence patters of erosion and sedimentation within the site and in the adjacent area.

The use of numerical modelling is essential to assess how breach design will alter the hydrodynamics at the sites. A survey conducted in 2013 indicates that although practitioners in the UK have great confidence in the hydrodynamic modelling tools used to assist project design, improvement is required concerning sediment processes (Esteves and Thomas 2014). At some sites, sediment dynamics responses following breaches can be complex, as observed at Freiston Shore (Lincolnshire, UK), where a managed realignment scheme was completed in 2002.

At Freiston Shore, the design of the managed realignment scheme was informed by modelling results and included: breaching of the flood defence (three breaches, each 50-m wide), upgrading of an existing embankment further inland and the creation of an artificial tidal creek system. During two years after breaching, development of the tidal creeks showed landward growth (through erosion of adjacent intertidal areas) at rates reaching 400 m/year or about 20 times greater than observed

---

[2] Information about the l'Aber Crozon was obtained from Brittany's General Inventory of Cultural Heritage, available from: http://patrimoine.region-bretagne.fr/sdx/sribzh/main.xsp?execute=show_document&id=MERIMEEIA29004203.

at natural conditions (Symonds and Collins 2007). A large volume of sediment was mobilised and subsequently deposited in adjacent areas destroying an oyster farm.[3]

The rapid erosion and deposition associated with the development of the tidal creeks was not anticipated. The enhanced sediment dynamics was attributed to the volume of water retained within the realignment site during the enhanced high spring tides which occurred few days after breaching. The flow of water out of the site and into the adjacent intertidal area persisted throughout the low tide until the next high water and exacerbated the existing drainage network overflowing the tidal creeks and causing the increased erosion (Symonds and Collins 2007).

Note that this 'FCA effect' was caused by the combination of the inland flood defence confining water within the site and the slower outflow restricted by the breaches. Freiston Shore is an example of managed realignment implemented through *realignment of defences* (as inland defences were upgraded as part of the scheme). However, similar effect could occur in projects implemented by breaching of defences if the realignment site is confined inland by a pre-existing embankment. It is therefore important to assess whether the relation between the volume capacity of the site and the drainage through the breaches and associated tidal creeks is adequate also during extreme high water events, to reduce potential negative impacts.

It is not evident from the available literature whether there has been an attempt to test whether models would be able to reproduce the observed hydrological-sediment dynamics response. Considering the availability of suitable data (e.g. Symonds and Collins 2007), this would be a relevant opportunity to validate model results and demonstrate that existing models are able to adequately support managed realignment design taking into account sediment dynamics.

3.4   Realignment of Defences

This method of implementation involves changing the position of the main line of defence to favour the development of a more dynamic coastline (Fig. 3.1d–f). Usually a new line of defence is built inland (or an existing one is upgraded) and the old defence is removed (Fig. 3.1d) or naturally or artificially breached (Fig. 3.1e). In the UK this has been the most common type of managed realignment, as it ensures a desirable level of protection to areas further inland. Projects including the construction of a new line of defence in anticipation of a high probability of old defences being naturally breached should also be considered as managed realignment. Projects involving seaward realignment of defences (Fig. 3.1f) with the objective of creating favourable conditions for the development of habitats may be considered within the broad definition of managed realignment (see Chap. 2). In The Netherlands, for example, the *Sand Motor* is a large beach nourishment schemes aiming to create a

---

[3] More information (including pre- and post-event photos) about this unexpected impact from the managed realignment at Freiston Shore through the perspective of the owner of the oyster farm is available at: http://washoyster.co.uk/.

Fig. 3.2 Realignment of the seawall at Littlehaven Beach in 2013 was an integral part of the local regeneration plan to increase the aesthetic and amenity value of the beach frontage, in addition to improvement of coastal protection (photo: Steve Burdett, courtesy of Royal HaskoningDHV)

buffer for storm impact and space for the development of back beach environments (see Chap. 5).

Creation of habitat is not always an objective for realignment of defences. Sometimes, due to continued coastal erosion or inadequate design/positioning, the alignment of coastal defences make them too exposed to the direct impact of waves (e.g. the embankment that was breached at Freiston Shore). High exposure reduces the life-time of the structure (e.g. undermining, structural damage) and the level of protection they offer (e.g. overtopping during storm surges). Realignment of defences may be implemented to reduce exposure and create opportunities for other benefits (e.g. recreation and amenity value). This was the case at Littlehaven Beach (South Tyneside, Northeast England). At Littlehaven beach the seawall was realigned (Fig. 3.2) switching from a protruding to a concave planform, to improve coastal defence, reduce maintenance costs and increase amenity value (Cooper et al. 2013).

3.5 Controlled Tidal Restoration

The main characteristics of controlled tidal restoration methods are: defences are maintained and tidal flow into the embanked area is controlled through culverts and sluices (Fig. 3.3). In controlled tidal restoration, realignment refers to the shoreline rather than the line of defence, as the high water line moves landwards within an embanked area. There are two types of controlled tidal restoration methods: regulated tidal exchange (RTE) and controlled reduced tide (CRT). These methods are

Fig. 3.3 (**a**) Location of the control reduced tide scheme at the Lippenbroek polder in the Schelde river, Belgium. (**b**) The polder is enclosed by dykes and tidal exchange occurs with water entering the site through the upper culvert and returning to the estuary through the lower culvert. (**c**) A view of the culvert system looking from inside the polder is shown in (modified from: Beauchard et al. 2011)

described in Chap. 2. RTE schemes are often implemented with the main objective of habitat restoration (e.g. Polder de Sébastopol, Fig. 2.2). CRT schemes aim to enhance creation of intertidal habitat within flood control areas (see Fig. 2.3); therefore they have both flood risk management and habitat creation as primary objectives. The use of flood control areas for agriculture is usually unsuitable due to the salinity of the estuarine water that enters the site through overtopping of flood defences during extreme high water level events. Therefore, these areas have become attractive for restoration of intertidal habitats to compensate for loss elsewhere.

In locations with high coastal development pressure, where land is scarce and expensive (e.g. Belgium), managed realignment implemented through the removal or breaching of flood defences is often unsuitable (Cox et al. 2006). Controlled tidal restoration through CRT offers opportunity for managed realignment in coastlines heavily engineered by flood defences and where flood risk mitigation is a serious concern. CRT schemes have been implemented in Belgium and the Netherlands, often as part of a transnational agreement for improving environmental quality in the Scheldt[4] estuary (see Chap. 4).

In terms of opportunities for habitat creation, controlled tidal restoration allows saltmarshes to develop on land with elevations considerably lower than it would be possible naturally (Vandenbruwaene et al. 2011). The characteristics of the tidal inundation and sedimentation patterns within RTE and CRT sites are considerably different than natural intertidal areas as the tidal flow is regulated by the size and elevation of sluices and culverts. Within RTE sites water level variations tends to be similar at neap and spring tides; while CRT schemes aim to create a neap-spring tidal variation with the wider range of inundation levels and frequency required for the establishment of the full spectrum of intertidal habitats (Maris et al. 2007; Jacobs et al. 2009).

---

[4] Scheldt is the English spelling for the Schelde, the original Dutch spelling.

On the other hand, the inflow and outflow of storm water at CRT schemes may increase the need of management when compared with RTE. The inflow of water during storms may enhance sediment accumulation, which needs to be controlled to avoid undesirable reduction of storage capacity; while emptying of storm water storage from the site may result in high erosion rates affecting habitat restoration (Cox et al. 2006). The Lippenbroek polder (Fig. 3.3) was an active agricultural area until 2003; construction of the CRT scheme started in 2004 culminating with the restoration of current tidal regime in March 2006 (Teuchies et al. 2012). The tidal amplitude within the CRT site is reduced from 5.2 m in the estuary to 0.9 m (Beauchard et al. 2011). A rapid sediment accumulation was observed, especially at lower elevations, where the agricultural soils were covered by ~30 cm estuarine sediments in three years (Vandenbruwaene et al. 2011).

Numerical modelling simulations indicate that differences in sedimentation patterns between natural and CRT marshes are likely to influence their long-term evolution (Vandenbruwaene et al. 2011). Results of these simulations show that, in conditions of sea level rise, CRT marshes may reach equilibrium at lower elevations and be less able to cope with rising water levels than natural marshes. However, the advantage of controlled tidal restoration is the opportunity for adaptive adjustments (e.g. modifying drainage systems and sluices) to accommodate for sea-level rise and other climatic changes.

3.6 Managed Retreat

The primary objective of managed retreat is to reduce the number of people, property and infrastructure at risk through planned retreat from hazard-prone areas. This method of managed realignment requires a long-term strategy for planning land-use changes that will create the space for the shoreline to respond dynamically to sea-level rise and erosion (Roca et al. 2008; Reisinger et al. 2014). Long-term and strategic planning makes managed retreat one of the few options available for developed coasts threatened by rising sea levels (Alexander et al. 2012). The implications of climate change on the management of coastal areas (including flood and erosion risk) is leading to an increased interest on managed retreat strategies at the local and national levels, as illustrated by the French national strategy for coastal management (see Sect. 4.8) and the removal of buildings in Texas (see Sect. 4.7) and Maui (Chap. 7).

In Europe, managed retreat experiences are rare, while elsewhere (e.g. the USA and New Zealand), this approach has been more widely discussed and implemented. Managed retreat is a long-term adaptive solution that promotes coastal sustainability and resilience in the face of climate change. It enables accruing social, economic and environmental benefits by: (a) reducing dependence on the increasing costs of providing protection through hard engineering; (b) reducing the number of people and property at risk from coastal defence failures during extreme events; (c) creating space for restoration or preservation of habitats; and (d) creating and enhancing recreational and amenity value (e.g. Reisinger et al. 2014).

Fig. 3.4 Cape Hatteras Lighthouse was moved landwards in 1999 from a position just behind the revetment seen in the foreground. This photo was taken on 16 July 2008 after landfall of hurricane Isabel (photo courtesy of The Program for the Study of Developed Shorelines, Western Carolina University)

In some cases, it involves the relocation of single structures at risk. The Cape Hatteras lighthouse and associated buildings (North Carolina, USA) were moved 880 m further inland in 1999 to protect the historic landmark from coastal erosion[5] (Fig. 3.4). Also in 1999 the Belle Tout lighthouse (East Sussex, UK), then converted to a private home, was moved 17 m away from the edge of the eroding cliff.[6] Although these examples may be technically and economically challenging, at the same time they are relatively simple as they were led by willing owners. The implementation of managed retreat in a larger-scale (e.g. involving entire communities) is more complex and requires political will (e.g. adequate policy and legislative measures), institutional capacity (e.g. integration between planning and management of flooding and erosion risk) and public engagement.

Legislation supporting managed retreat usually involves alterations to private property rights based on defined thresholds of risk. It might include, for example, restrictions to restoration or reconstruction of properties affected by flooding or erosion if they are within high risk zones. In the aftermath of the Xynthia storm of 2010, a compulsory purchasing of property at high risk was adopted in France.

[5] Photos and more information about the relocation of Cape Hatteras lighthouse are available at: http://www.nps.gov/caha/historyculture/movingthelighthouse.htm.

[6] A time-line and images illustrating the history of the Belle Tout is available at: http://www.belletout.co.uk/history.html.

In USA financial assistance to facilitate relocation if a property is considered at threat of imminent damage from erosion or flooding was supported through the Upton-Jones amendment adopted in 1988. The amendment was repealed in 1994 due to lack of interest from the public. However, there has been some appeals for the amendment to be reinstated (e.g. Nags Head in North Carolina, USA). Similar initiatives exist at the state level, such as in Texas where the government offers reimbursement of up to US$ 50,000 to cover for landowners costs related to moving out from the public beach (see Sect. 4.7).

Public opposition can delay or prevent managed realignment projects and therefore early engagement with communities and stakeholders is essential (see Chap. 8). Public support for managed retreat depends on who will pay the costs, how the loss in assets and changes in existing use rights are managed and who benefits from a natural shoreline retreat (e.g. Reisinger et al. 2014), whether alternative land is available and the underpinning cultural values of communities and individuals (Alexander et al. 2012).

Managed retreat usually requires strong integration between long-term planning and the sustainability of risk reduction measures, which is often deficient in public administrations. However, challenging times require drastic changes and the only safe climate-proof response at all temporal and spatial scales is to reduce the number of people and assets at risk. As it is an effective mechanism to reduce risk from both climatic variability and extreme events, managed retreat is increasingly been implemented (or planned) in many locations worldwide.

However, implementing managed retreat is complex due to the variety of social and political issues it involves. The delivery of long-term strategies is usually hindered by inadequate institutional systems and public resistance, which are accustomed to short-term remediation that can be immediately claimed (e.g. within the 4–5 years of administrative cycles) and the inability to understand that compromising personal or sectorial gains is needed so greater social, environmental and economic benefits are realised.

References

Alexander, K. S., Ryan, A., & Measham, T. G. (2012). Managed retreat of coastal communities: Understanding responses to projected sea level rise. *Journal of Environmental Planning and Management, 55*(4), 409–433.

Beauchard, O., Jacobs, S., Cox, T. J. S., Maris, T., Vrebos, D., Van Braeckel, A., & Meire, P. (2011). A new technique for tidal habitat restoration: Evaluation of its hydrological potentials. *Ecological Engineering, 37,* 1849–1858.

Cooper, N. J., Wilson, S., & Hanson, T. (2013). *Realignment of Littlehaven Sea Wall, South Tyneside, UK.* Proceedings of the Coasts, Marine Structures and Breakwaters Conference (Edinburgh, 18-20 September 2013), Institution of Civil Engineers, London 10 p. http://www.ice.org.uk/ICE_Web_Portal/media/Events/Breakwaters%202013/Realignment-of-Littlehaven-Sea-Wall,-South-Tyneside,-UK.pdf. Accessed 1 Feb 2014.

Cox, T., Maris, T., De Vleeschauwer, P., De Mulder, T., Soetaert, K., & Meire, P. (2006). Flood control areas as an opportunity to restore estuarine habitat. *Ecological Engineering, 28,* 55–63.

Esteves, L. S., & Thomas, K. (2014). Managed realignment in practice in the UK: Results from two independent surveys. *Journal of Coastal Research, Special Issue 70,* 407–413.

Jackson, N. L., & Nordstrom, K. F. (2013). Removing shore protection structures to facilitate migration of landforms and habitats on the bayside of a barrier spit. *Geomorphology, 199,* 179–191.

Jacobs, S., Beauchard, O., Struyf, E., Cox, T., Maris, T., & Meire, P. (2009). Restoration of tidal freshwater vegetation using controlled reduced tide (CRT) along the Schelde Estuary (Belgium). *Estuarine, Coastal and Shelf Science, 85*(3), 368–376.

Maris, T., Cox, T., Temmerman, S., De Vleeschauwer, P., Van Damme, S., De Mulder, T., Van den Bergh, E., & Meire, P. (2007). Tuning the tide: creating ecological conditions for tidal marsh development in a flood control area. *Hydrobiologia 588,* 31–43.

Nottage, A., & Robertson, P. (2005). *The saltmarsh creation handbook: A project manager's guide to the creation of saltmarsh and intertidal mudflat.* London: RSPB.

Reisinger, A., Lawrence, J., Hart, G., & Chapman, R. (2014). From coping to resilience: The role of managed retreat in highly developed coastal regions of New Zealand. In B. Glavovic, R. Kaye, M. Kelly, & A. Travers (Eds.), *Climate change and the coast: Building resilient communities.* London: CRC Press.

Roca, E., Gamboa, G., & Tàbara, J. D. (2008). Assessing the multidimensionality of coastal erosion risks: Public participation and multicriteria analysis in a Mediterranean Coastal System. *Risk Analysis, 28*(2), 399–412.

Symonds, A. M., & Collins, M. B. (2007). The establishment and degeneration of a temporary creek system in response to managed coastal realignment: The Wash, UK. *Earth Surface Processes and Landforms, 32*(12), 1783–1796.

Teuchies, J., Beauchard, O., Jacobs, S., & Meire, P. (2012). Evolution of sediment metal concentrations in a tidal marsh restoration project. *Science of the Total Environment, 419*(1), 187–195.

Vandenbruwaene, W., Maris, T., Cox, T. J. S., Cahoon, D. R., Meire, P., & Temmerman, S. (2011). Sedimentation and response to sea-level rise of a restored marsh with reduced tidal exchange: Comparison with a natural tidal marsh. *Geomorphology, 130,* 115–126.

Chapter 4
Examples of Relevant Strategies and Policies

Luciana S. Esteves

Abstract Managed realignment projects are increasingly popular in the western developed countries. The underpinning drivers for the implementation of managed realignment are generally similar and revolve around restoration of natural habitats, improved flood and erosion risk management and climate change adaptation. Emerging national and regional strategies will play an important role in the increase of managed realignment approaches being implemented in the next decades. This chapter provides eight examples of strategies adopted by different countries or regions that are likely to influence the dissemination of managed realignment at present and in the future. Examples are drawn from the UK, Belgium, the Netherlands, Germany, France and the USA. The strategies are discussed in terms of their primary objectives, scope and the preferred mechanisms of implementation.

4.1 Geographical Background

Independently on the terminology used nationally or locally, managed realignment projects are increasingly popular in the western developed countries. The underpinning drivers for the implementation of managed realignment are generally similar and revolve around restoration of natural habitats, improved flood and erosion risk management and climate change adaptation. Emerging national and regional strategies will play an important role in the increase of managed realignment approaches being implemented in the next decades. However, there are differences in the strategies adopted by different countries, their primary objectives and the preferred mechanisms of implementation. The multiple functions and benefits provided by managed realignment are widely recognised; but some countries have placed stronger emphasis on the environmental benefits while others primarily focus on sustainable management of flood and erosion risk in the face of climate change.

L. S. Esteves (✉)
Faculty of Science and Technology, Bournemouth University, Talbot Campus,
Poole, Dorset, BH12 5BB, UK
e-mail: lesteves@bournemouth.ac.uk

L. S. Esteves, *Managed Realignment: A Viable Long-Term Coastal Management Strategy?*,   45
SpringerBriefs in Environmental Science, DOI 10.1007/978-94-017-9029-1_4,

    In the UK managed realignment is commonly implemented through realignment of defences or controlled tidal restoration (see Appendix) with nature conservation taking the spotlight. In Belgium and the Netherlands, flood protection is a primary concern and projects focus on adding environmental benefits to new flood defence strategies. In the USA, implementation of managed realignment is more diverse, in some areas focusing on managed retreat and others on breach, removal or realignment of defences focusing primarily on nature restoration. Although experiences are still few and scattered, the interest in managed retreat is also increasing in other countries (e.g. Australia and New Zealand).

    The first managed realignment projects in Europe were implemented in France in 1981 and in Germany in 1982 and the Netherlands in 1989 (see Appendix). These were isolated initiatives to attend to local needs. Currently, projects tend to be designed and implemented as part of regional or estuary-wide strategies, some of which are described in this chapter. The UK is said to have the highest number of managed realignment projects implemented to date. However, this might not be the case if all mechanisms of implementation are considered. It would be necessary to have an inventory of all managed realignment projects implemented in the USA (which might include some of the projects of wetland restoration, hydraulic restoration and managed retreat), for example, to allow adequate comparison. This is a difficult task because of the range of terminology used and the many different programmes and individual projects of relevance spread along different countries and regions.

    The Online Managed Realignment Guide (OMReG,[1] ABPmer 2013) provides a list of projects implemented in Europe, including an overview of their main characteristics and location. An updated list of managed realignment projects in Europe, based on information obtained from a range of sources, including verified projects listed in the OMReG, is provided in the Appendix. Projects are listed alphabetically per country and information is provided about the mechanism of implementation, date, size of the realigned site (in hectares) and relevant sources of information.

    Due to the range of terminology used in different countries, information is often scattered and therefore this list should be considered a work in progress. No attempt has been made here to list schemes implemented outside Europe due to the time and effort this would require. However, a few projects implemented in the USA are listed to illustrate that some projects, although not disseminated as managed realignment, do fit the definition suggested in this book. Therefore, managed realignment experiences worldwide are actually more widely spread than first anticipated.

    The list indicates that the number of managed realignment projects is increasing through time. Not only many large-scale projects are currently under construction, many more are planned to be implemented in the next decades as part of national and regional programmes. As a consequence of these regional programmes, it is clear that managed realignment schemes so far tend to be implemented in certain geographical areas. Many projects are planned in the Scheldt estuary (Belgium and Netherlands) and in the UK many schemes have been implemented in the Blackwater and Humber estuaries (Figure 4.1). To illustrate the geographical differences and

---

[1] OMReG is planned to be re-structured in the near future and re-launched as the Online Marine Register to include other types of marine habitat creation projects (as announced by Susanne Armstrong during the ABPmer Habitat Creation Conference in November 2013, London).

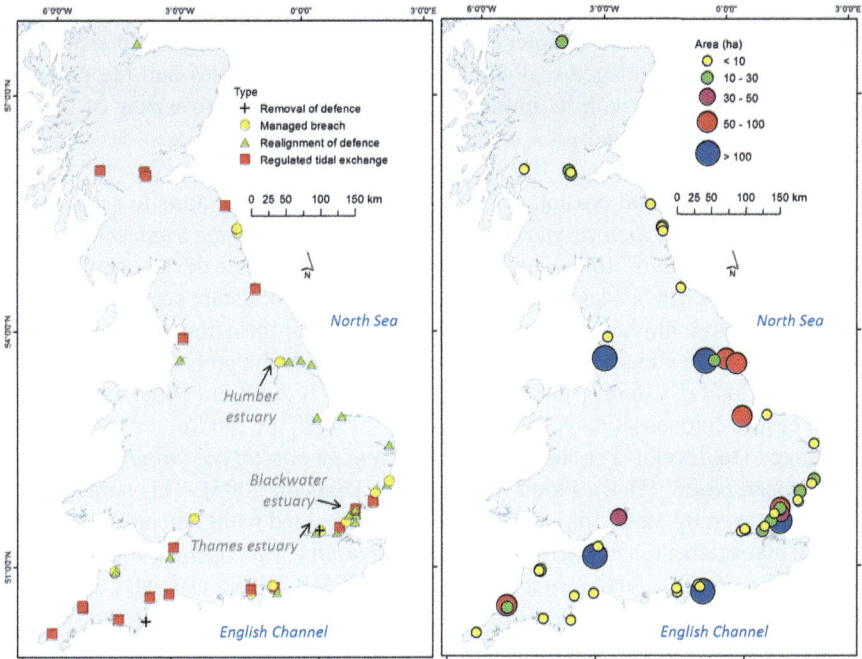

Fig. 4.1 Distribution of the 54 managed realignment projects implemented in the UK (up to February 2014) showing (**left**) the type of scheme and (**right**) the size of the realigned area

similarities, some of the main policies and regional plans driving the implementation of managed realignment are described here. These should be taken as examples of the strategies that many governments at different levels (local, regional, national, transnational) are now adopting towards more sustainable coastal management and climate change adaptation.

4.2   Making Space for Water and Making Space for Nature, UK

In the UK, especially in England, managed realignment is implemented to create intertidal habitat and to deliver more sustainable flood risk management, e.g. by reducing costs and aggregating environmental and amenity values. The Department for Environment Food and Rural Affairs (Defra) is responsible for policy-making concerning nature conservation and flood risk. The Environment Agency is the operating authority responsible for implementing policy related to coastal erosion and flood management. Under the *Flood and Water Management Act*[2] 2010, local

---

[2] The roles and responsibilities for flood risk management in England and Wales are summarised at: http://www.local.gov.uk/local-flood-risk-management/-/journal_content/56/10180/3572186/ARTICLE. The full text of this legislation is found at: http://www.legislation.gov.uk/ukpga/2010/29/contents.

authorities are the designated Lead Local Flood Authority, which have the duty for managing local flood risk. Chapter 8 presents the Environment Agency perspectives on the drivers for managed realignment in England and Wales and the practical challenges associated with its implementation. Here, a brief overview of the national policies is presented.

In 2005, *Making Space for Water* was implemented as the new strategy for managing flood and coastal erosion risk in England and Wales. It aims to *"reduce the threat to people and their property"* and *"deliver the greatest environmental, social and economic benefits"* following the principles of sustainable development (Defra 2005). Climate change adaptation and the statutory duty of nature conservation (and related economic implications) are the main drivers for this strategy (see Chaps. 8 and 9). The implementation of managed realignment is the preferred approach for managing flood risk in rural areas and to create habitat to offset or compensate loss.

To ensure compensation is delivered when required, the Environment Agency is encouraged to develop a strategic plan to *"anticipate habitat creation requirements and opportunities"* (Defra Flood Management Division 2005). This strategic approach is taken by identifying opportunities for managed realignment into existing coastal management instruments, such as the *Shoreline Management Plans*[3] (SMP) and the *Catchment Flood Management Plans*[4] (CFMP). Defra estimates that about 100 ha of intertidal habitat needs to be created per year to compensate loss due to coastal squeeze and development projects within Natura 2000 sites especially in south and east England. Taking into consideration Defra's regional estimates on the extent and type of habitat that needs to be created, compensation may be delivered through habitat creation programmes (e.g. the Anglian Regional Habitat Creation Programme).

Not surprisingly managed realignment has been a preferred coastal management strategy in England in the twenty-first century. The Appendix lists 54 managed realignment projects implemented in the UK since 1991, their geographical distribution, type of implementation method and size of the realigned area can be seen in Fig. 4.1. The shift from a flood risk management strategy based on hard engineering to the more environmentally-focused managed realignment was a core aim of the *Making Space for Water* policy. In relative terms, this shift in flood management in England can be compared to the 1960s shift from hard engineering to beach nourishment in the management of coastal erosion in the USA.

The UK's National Adaptation Programme (Defra 2013) incorporates land use changes as a mechanism to enhance climate and flood control and sets out the development of *"a strategic plan for coastal realignment"* as one of the actions to address priority risks. The government plans to realign, in England and Wales, a

---

[3] SMPs are non-statutory guidance on coastal management alternatives. Currently, 22 SPMs cover the entire coastal length of England and Wales and each identify a management plan in three time horizons (0–20, 20–50 and 50–100 years). The SPMs can be accessed from: http://www.environment-agency.gov.uk/research/planning/105014.aspx.

[4] CFMPs describe flood risk from all sources (except the sea) within the catchment and recommend management approaches over the next 50–100 years. They can be accessed from: http://www.environment-agency.gov.uk/research/planning/33586.aspx.

total of 111 km by 2016 and 550 km (10 % of the coastline) by 2030 resulting in the creation of 6,200 ha of intertidal habitat at a cost of £ 10–15 million per year (Committee on Climate Change 2013). Until November 2013, about 66 km of England's coastline have been realigned (Committee on Climate Change 2013). Therefore, the government plans require that the length of realigned shorelines almost double in the next two years (2014–2016) and increase eight-fold in 16 years (2014–2030).

The independent review *Making Space for Nature* (Lawton et al. 2010) assessed the sustainability of natural environments in England in the face of climate change. The report makes 24 recommendations, which will set the direction for upcoming habitat restoration and re-creation strategies. It identifies that natural habitat sites are too small and isolated and are not able to maintain provision of all ecosystem functions, especially when climate change impacts on existing sites are considered. A more effective ecological network is required if society is to benefit from ecosystem services of water-quality, flood and erosion control and carbon storage. Chapter 8 describes the implications of this report for the implementation of managed realignment.

This ambitious strategy illustrates the importance of managed realignment to deliver multiple functions to satisfy the need to adapt to climate change, compensate for habitat loss and provide sustainable coastal protection. Implementation of managed realignment at the scale and rate planned by the government requires: (a) ensuring that the identified sites have favourable conditions to the development of the habitats to be created; (b) the willingness of landowners; and (c) increasing contribution from external funding. These three factors are known to delay or hinder managed realignment projects (see Chap. 8).

4.3 Sigma Plan, Belgium

Belgium's coastline is heavily engineered by flood defences and flood risk mitigation is a serious concern. It is a country with high coastal development pressure where land is scarce and expensive (Cox et al. 2006). As in other European countries, restoration of intertidal habitats is required to abide to environmental law. The need to provide improved protection against floods and nature restoration has led to the implementation of managed realignment, mainly through controlled reduced tide (CRT). Most of CRT projects (13 out of 15) in Belgium have been implemented along the Scheldt estuary and more schemes are planned to be implemented by 2030.

The Scheldt marks the border between the Netherlands and Belgium. The Scheldt Estuary Development Plan, a Netherlands-Flanders agreement, establishes an integrated long-term vision for the estuary's accessibility, environmental conservation and flood protection. Human interference (e.g. dredging and land reclamation) in the Scheldt estuary has reduced intertidal habitat to less than 50 % over the past century (Meire et al. 2005). The economic and ecological importance of the Scheldt has caused historic cross-border conflicts; more recently involving a case of managed realignment (see Sect. 4.4).

In Belgium, the Sigma Plan[5] is a major regional strategy aiming to deliver improved flood risk management in the Flanders region and enhancement of environmental conditions along the Scheldt and its tributaries. It was developed as a response to a major flood that in 1976 badly affected Antwerp and other parts of Flanders. The Sigma Plan included three main measures: 512 km of enhanced dikes, flood control areas and one flood barrier (which was considered unfeasible and dropped from the plan). In 2005 the Sigma Plan was revised to deliver a more integrated plan that attends to current multifunctional needs (i.e. recreation, nature restoration, climate change adaptation and sustainability of economic activities). A good overview of the Sigma Plan is provided in Beukelaer-Dossche and Decleyre (2013).

Managed realignment plays a key role in the revised Plan, which includes about 1,300 ha of flood storage area and total over 850 ha of new habitat. The Sigma Plan has been developed by the Waterways and Sea Canal Agency (which is responsible for flood protection and navigation) in partnership with the Agency for Nature and Forest. A number of large scale projects have been identified, including realignment of flood defences and CRT, and systematically planned to be implemented in time-periods of five years, with the final delivery of the plan expected in 2030. The projects that have already started are included in the list of management realignment projects in Europe (see the Appendix). Lippenbroek (Fig. 3.3) was the first CRT project piloted in Belgium (Jacobs et al. 2009; Beauchard et al. 2011; Teuchies et al. 2012).

To reduce impact of the revised Sigma Plan on urban and agricultural areas, it was decided to concentrate most of the projects into preferred areas. Although affecting a smaller area, the impact on the value of farmland in the affected areas is considerable and the government decided for a policy of expropriation and freehold purchasing that is currently being tested in the Kalkense Meersen (Van Rompaey and Decleyre 2013). Other measures to facilitate land purchase include: creation of a land bank (so land is available to offer as exchange to owners affected by the Sigma Plan); € 2,000/ha in financial incentives (above market value) for willingness to sell; relocation; compensation for loss of production; low interest loans etc. Another set of measures are also in place to enhance the recreational value of the projects.

4.4   Room for River and Building with Nature, The Netherlands

The Netherlands is a country known for reclaiming land from the sea and engineering coastlines to ensure flood protection and secure the functioning of the country (see Chap. 5). The importance of the dams-dikes system and a robust coastal management plan cannot be understated in a country where about two thirds of its area is below sea level (Brouwer and van Ek 2004). Rijkswaterstaat (Ministry of

---

[5] Information about the Sigma Plan and its projects can be found at: http://www.sigmaplan.be/en/sigma-plan.

Infrastructure and Environment) is responsible for the main waterways, water systems and road networks in the Netherlands, including flood protection.

The catastrophic impact of the 1953 floods reinforced an already strong culture of engineering protection of polders. In the aftermath of this major event, the Delta Plan was devised to enhance protection of the country against high water levels through the constructions of dams. In 1958 the first of the 'deltaworks' was concluded; as a result of hard engineering conducted under the Delta Plan the Dutch shoreline was shortened by 700 km. In 1990, the government implemented a new policy known as *Dynamic Preservation of the Coastline*, stating that the coastline should not be allowed to retreat inland of its 1990 position (Hillen and Roelse 1995). Beach nourishment became the main mechanism of implementation.

Climate change adaptation and environmental concerns have led to a new policy development, the *Delta Programme*,[6] which is expected to be completed between 2015 and 2020. The *Delta Programme* aims to integrate flood protection, freshwater availability and spatial planning and it has nine sub-programmes (three thematic and six geographic): Safety; Freshwater; New Urban Developments and Restructuring; Ijsselmeer region; Rhine Estuary-Drechtsteden; Southwest Delta; Rivers; Coast and Wadden region.

The Dutch policy has evolved through time to respond to new challenges and the current Delta Programme not only recognises the importance of naturally evolving coasts but also is underpinned by concepts of eco-engineering and building with nature (Rijkswaterstaat and Deltares 2013). Currently eco-engineering approaches that promote the multiple benefits (e.g. ecosystem services) provided by the creation of natural environments such as dunes and wetlands are increasingly used in the Netherlands.

Many of the eco-engineering projects described as *building with nature* (e.g. Rijkswaterstaat and Deltares 2013) could be classified as managed realignment. However, the perception of 'retreat' associated with the term 'managed realignment' restricts its use and acceptance in the Netherlands (Eertman et al. 2002). Chapter 5 describes the Dutch experience concerning managed realignment and places it in the wider context of flood and erosion protection. To date, ten managed realignment projects have been implemented in the Netherlands, including one of the oldest schemes in Europe, implemented in 1989 at Holwerder Zomerpolder.

4.5 Cross-Border Conflict in the Scheldt Estuary

The Scheldt estuary is about 160 km long, and the mean high water line has risen faster along the inner estuary (15 mm a^{-1}) than at its mouth (3 mm a^{-1}) in the last 100 years (Temmerman et al. 2004). Due to high flood risk, the estuary is constrained by a network of flood defences.

A common agreement between the Flemish and Dutch governments involves mutual cooperation on the management for the safety, environmental quality and

---

[6] The latest developments and details of the new policy can be found at: http://www.government. nl/issues/water-management/delta-programme.

accessibility along the estuary (De Beukelaer-Dossche and Decleyre 2013). Among the agreed objectives were the number of times the navigation channel would be deepened and the creation of 600 ha of low-dynamic estuary environment (300 ha in the saltwater part of the estuary and 300 ha in the brackish part of the estuary). The mechanisms to be adopted by each government to deliver the long-term plan were identified in the agreement (e.g. the deepening of the channel given access to the Port of Antwerp, a revision of the Sigma Plan and habitat restoration projects). The Dutch government has changed the plans concerning some of the habitat restoration projects proposed through managed realignment, creating conflict with the Flemish government.

In 1998, Rijkswaterstaat proposed the implementation of managed realignment to compensate the loss of intertidal habitats within a designated Special Protection Area in the Scheldt Estuary caused by the dredging of the navigation channel. Local communities opposed the idea due to concerns about increasing flood risk; thus, leading local authorities to reject the proposal, which the Dutch government withdrawn in 2000. This change in plans caused havoc across the border as in Belgium it was perceived as a break in the bilateral agreement.

A similar issue has arisen more recently and the case was taken to the European Parliament. Under the EU Habitats Directive member states must take appropriate measures to restore favourable conservation status and to avoid further deterioration of Natura 2000 sites. According to the European Commission, it is the competence of the country responsible for the activity causing environmental damage to identify the most suitable measures, the location where they will be implemented and the time-frame in which actions will be taken to avoid further deterioration.

In 2005 the Netherlands identified that restoration of tidal flow into the Hedwige Polder was the most suitable measure to compensate for damage caused by the deepening of the Western Scheldt channel. The area is at the border between Belgium and the Netherlands, with the Hedwige polder being on the Dutch side and the Prosper polder on the Belgium side. In June 2011, the Dutch authorities communicated to the European Commission that they would adopt alternative measures instead of creating intertidal habitats in the Hedwige polder.

Some of the alternative measures included creation of habitat in an area that was already planned to compensate loss caused elsewhere. This double accounting was not well received by the Flemish authorities, which have filed three separate sets of questioning[7] to the European Parliament. The main concerns were: (a) political motivation; (b) the time-frame in which the alternative compensation measures were proposed (i.e. most actions were proposed to take place after 2015); and (c) the equivalence of the quantity and quality of the habitat that could be created at a site under hydrodynamic different conditions.

---

[7] Parliamentary questions E-006402/11(30 June 2011), E-006507/11 (4 July 2011) and P-006822/11 (12 July 2011).

The European Commission provided a consolidated response to the three set of questions on 15th September 2011.[8] The response was based on three main points:

1. The issue revolves on restoration of nature's favourable conservation status and the Commission is not required to provide a formal opinion on the measures elected by the competent authorities of the Netherlands.
2. The Habitats Directive does not establish a precise time-frame in which favourable conservation status of designated species and habitats should be achieved. However, it is expected that the Netherlands take timely and efficient measures to avoid irreversible ecosystem deterioration in the Western Scheldt.
3. As the decision of the Dutch Government on alternative compensation measures created controversy, the Commission requested that the effectiveness of the new proposal and its equivalence to the original plan should be substantiated scientifically to remove any reasonable questions.

The realignment of defences at the Hedwige-Prosper polder has being finally agreed and works on both sides of the border have started in preparation to restore tidal flow into 465 ha of previously reclaimed land.

4.6   Cities and Climate Change, Germany

Managed realignment in Germany has been implemented along the Baltic and North Sea coasts and along main rivers, such as the Elbe and the Weser. A total of 29 managed realignment projects have been implemented in Germany (see list in the Appendix), most involving compensation for habitat loss (Armstrong 2013). Coastal managed realignment sites are found in Lower Saxony (by the North Sea) and Mecklenburg–Western Pomerania (by the Baltic Sea), but the objectives differ across these two areas (Rupp-Armstrong and Nicholls 2007). In Lower Saxony, managed realignment is usually implemented for compensation reasons (i.e. loss of intertidal habitats due to coastal development, port construction etc.). In western Pomerania managed realignment often combines the need for improvement of flood defences and creation of new intertidal habitats.

The coastal managed realignment projects tend to be of larger size than the estuarine sites. However, the concept of multiple functions is more evidenced in recent estuarine projects. For example, managed realignment on the Elbe island of Wilhelmsburg (called Pilot Project Kreetsand[9]) is part of the project Elbe Islands

---

[8] OJ C 128 E, 03/05/2012: http://www.europarl.europa.eu/sides/getAllAnswers.do?reference=P-2011–006822&language=EN.

[9] A description of the project and relevant links are available at: http://www.iba-hamburg.de/en/themes-projects/elbe-islands-dyke-park/pilot-project-kreetsand/projekt/pilot-project-kreetsand.html.

Fig. 4.2 Aerial view of the Elbe river and the location of the Pilot Project Kreetsand at the east margin of the Wilhelmsburg island. Note that the image shows a visualisation of how the realigned site will look when implemented (project completion is expected in 2015). The realignment of defences that confine the Kreetsand tidal space was completed in 1999 (Image courtesy of HPA/ Studio Urbane Landschaften)

Dyke Park (Deichpark-Elbinsel[10]), which aims to combine the creation of flood storage space with provision of a recreational park. These initiatives are part of the *Cities and Climate Change* programme promoted by IBA Hamburg, which include a series of projects concerning 'climate-compatible' sustainable development for the metropolitan Hamburg area.

The Pilot Project Kreetsand is being developed by the Hamburg Port Authority and IBA Hamburg and involves breaching of existing defences at the margin of the river Elbe (Fig. 4.2). The objective is to allow flooding of 30 ha to create a tidal space of around 1 million m^3. Increasing space for tidal dynamics in the Elbe is vital for the future viability of the Hamburg Port, which will benefit from reduced sedimentation and flood risk. The project is associated with design and landscape features aiming to promote nature conservation and recreational value.

---

[10] The process of planning and implementation of the Elbe Islands Dyke Park, including stake-holders engagement and the landscape works to incorporate multiple functions to existing flood defences, is described at: http://www.osp-urbanelandschaften.de/fileadmin/user_upload/osp_ur-banelandschaften_Deichpark_Elbinsel_english.pdf.

4.7 Coastal Wetlands Planning, Protection and Restoration Act, Louisiana, USA

The Coastal Wetlands Planning, Protection and Restoration Act (CWPPRA) was enacted in 1990 to assist projects aiming to reduce loss and restore coastal wetlands in the USA. CWPPRA annual budget has ranged between US$ 30–80 million. The program aims to: integrate agencies to steer operations; plan and implement restoration projects; and maintain a comprehensive monitoring program to evaluate performance. Strategic plans to address wetland loss in Louisiana have been developed and managed by a partnership between the State of Louisiana and five Federal agencies (Army Corps of Engineers, Environmental Protection Agency, Fish and Wildlife Service, Natural Resources Conservation Service, National Marine Fisheries Service and U.S. Geological Survey/National Wetlands Research Center), which form the CWPPRA Task Force.

The first project was completed in April 1994 (LaBranche Wetlands) and a total of 151 projects covering more than 44,500 ha have been funded since.[11] A range of techniques is used in the CWPPRA funded projects, named: barrier island restoration (natural coastal protection); marsh creation (offsetting habitat loss); shoreline protection (hold the line); hydrological restoration (improving stability); terracing (innovation); freshwater and sediment diversion (restoring natural processes).[12] Although named using different terminology, some of the projects and techniques used fit within the broad definition of managed realignment. Restoration of tidal flow into impounded areas by installing culverts or other methods is a form of *controlled tidal restoration*. Breaching the Mississippi river embankments to allow distribution of sediment to feed the wetlands can be considered *managed realignment through breach of flood defences*. Depending on implementation methods and objectives, some of the projects of barrier islands restoration that aim to provide shelter for the development of back barrier wetlands may be consider a form of *realignment of defences*.

The selection of projects starts every January with new applications from local communities, agencies and government. A workgroup will then short-list a number of candidate projects that will be evaluated for feasibility and cost-effectiveness. A budget is estimated to implement, maintain and operate the project for 20 years and the CWPPRA Task Force will select the projects that will be funded according to a set of criteria. Demonstration projects are also selected to test new methods and applications that might be able to improve efficiency and results. One of the strengths of the programme is the systematic monitoring and the online dissemination of data publicly available online. This monitoring and data sharing is the basis of the 'adaptive management', which refers to the application of lessons learned to improve existing and future individual projects and the overall achievements of the programme.

---

[11] A list of CWPPRA projects in Louisiana is available from: http://lacoast.gov/new/Projects/List.aspx.

[12] The brochure *Partners in Restoration—20th Anniversary Portfolio*, produced by the CWPPRA Outreach Committee and the U.S. Geological Survey National Wetlands Research Center describes the range of techniques and provides background information about the work of CWPPRA. Available at: http://lacoast.gov/products/Portfolio_of_Success_Final_web.pdf.

In 2002, the CWPPRA projects were reviewed to identify achievements and actions needed to improve existing and future projects (Raynie and Visser 2002). The outcomes of the review indicate that evaluation of achievements is impaired when the original intent of the project is lost or becomes secondary as the project evolves. It is important therefore that success criteria and targets are clearly defined at the planning stage and monitoring should evaluate whether progress is being achieved towards the set target so adjustments can be made if required.

4.8   Texas Open Beaches Act, USA

The Texas Open Beaches Act (OBA), enacted in 1959, establishes that the public has free and unrestricted access to the State's beaches,[13] commonly known as the 'wet beach', or the area between the mean low tide and the mean high tide. The 'dry beach' between the mean high tide and the natural vegetation line can be privately owned but may be subjected to public beach easement (i.e. right of way). Under the OBA, structures on the public beach easement are prohibited if posing hazard to safety or interfering with free and unrestricted public access.

About 63 % of Texas shorelines are critically eroding (Patterson 2007); through time erosion results in structures being located within the public beach easement. In 2006, the 'structure removal initiative' was created to reduce litigation between the State and property owners in these situations. Through funding from the Coastal Erosion Planning and Response Act (CEFRA), the government offers up to US$ 50,000 to reimburse property owners for the costs to move structures located within the public beach easement to a suitable, landward location (Fig. 4.3). Therefore, the 'structure removal initiative' is a mechanism to support the implementation of managed retreat.

According to the Texas General Land Office, 16 structures have been relocated without litigation since this initiative was implemented. CEPRA also encourages local governments to establish setback lines, restore and protect dunes and purchase properties seaward of setback lines. The objective is to improve public access to the beach and reduce risk to people and property during storms.

In 2012 the Texas Supreme Court ruled that the location of the public beach easement should not be affected by sudden erosion events (e.g. due to storms or hurricanes); but may move due to gradual changes resulting from coastal dynamic conditions. This decision was the outcome of a lawsuit[14] filed by a property owner against the Texas Land Commissioner over the interpretation of the OBA. The complainant owned several houses in Galveston's West Beach, which were located in the public beach easement as a result from the impact of Hurricane Rita in 2005. As customary in these cases, the General Land Office sent a letter to the property owner stating that the houses were subject to removal under the Open Beaches Act.

---

[13] A description of the Act and how it is enforced is provided by the Texas State General Land Office at: http://www.glo.texas.gov/what-we-do/caring-for-the-coast/open-beaches/.

[14] Severance v. Patterson, 370 SW 3d 705– Tex: Supreme Court 2012.

Fig. 4.3 Example of house relocation under the Texas Open Beach Act. The photos on the top show the beach before (**left**) and after relocation (**right**). The house was located within the public beach easement at Sunny Beach, West Galveston Island and was relocated further inland (*bottom*) on 1 Nov 2010. (photos courtesy of Texas General Land Office)

The consequences of the Supreme Court decision were: (1) some of the beach houses under litigation were purchased by the Federal Emergency Management Agency (paid by taxpayers) so they could be removed; and (2) the OBA has been amended (by the H.B. 3459 which took effect on 1 September 2013) to allow temporary suspension of the line of vegetation as the landward limit of the public beach easement when it is altered by a meteorological event.

4.9   National Strategy for the Integrated Management of the Shoreline, France

Despite the fact that the first managed realignment project in Europe has been implemented in France (Aber de Crozon in 1981), the idea of managed realignment in this country is still at its inception. The *Conservatoire du Littoral*[15] is currently the main player, implementing managed realignment in some of its coastal properties. In 2012, the *Conservatoire du Littoral* published a booklet about coastal management options

---

[15] It is a government organisation created in 1975 to ensure the protection and management of coasts, estuaries and other water bodies. Land acquisition and subsequent creation of natural reserves is one of the mechanisms adopted to improve sustainable use and conservation of the natural environment. On the time of writing, the *Conservatoire du Littoral* owns 153,288 ha across France (http://www.conservatoire-du-littoral.fr/).

in the face of climate change in which it advocates land use changes, realignment of defences and managed retreat strategies, as long-term and sustainable options.

This approach reflects the direction set by the *National Strategy for Integrated Coastal Zone Management* proposed in 2011[16], which explicitly indicates the need to plan for the retreat of coastal activities and structures as one of its principles. The programme of actions (Ministère de l'Écologie, du Développement durable et de l'Énergie 2012) is based on four priorities and nine actions, one of the priorities involves the re-organisation of the territorial space and action 7 refers to the provision of mechanisms for the implementation of relocation strategies. Other principles include: prohibit occupation in high risk areas; integration of spatial planning and reduction of risk from natural hazards; protection and restoration of coastal ecosystems; and planning for climate and environmental change at time-scales of 10, 40 and 90 years.

Existing managed realignment experiences in France are still limited and include realignment of defences at Aber de Crozon and Île Nouvelle, controlled tidal restoration at the Polder de Sébastopol (Fig. 2.2), and retreat of a road and other land uses in the Languedoc-Roussillon coast (between Sète and Marseillan). Lack of awareness and poor acceptance of coastal communities and planners is still a major constraint to the implementation of managed realignment (Bawedin 2004; Goeldner-Gianella 2007; SOGREAH 2011).

Interest however is arising on the subject and studies have been commissioned on the feasibility of managed realignment at few sites. For example, in 2011 a committee has been established to assess the feasibility of managed realignment at Bas-Champs de Cayeux (Picardy). This reclaimed land is protected by embankments, which have been breached and flooded during storms in the past. About € 55 million have been spent between 1971 and 2014 to protect this area.

The discussions with local communities in the Bas-Champs area revolve around the relative comparison between the consequences of flooding due to an accidental breach and the planned restoration of tidal inundation in the area. However, such discussion is not new in the area and studies have addressed the subject since the accidental breach of 1990 caused severe flooding in the area (e.g. Bawedin 2000, 2004). A social survey with 250 local residents indicated that 69 % of respondents oppose to managed realignment at Bas-Champs and only 10 % accepts it unreserved (SOGREAH 2011). The respondents that would accept some form of realignment, 41 % would prefer it to occur through an accidental breach of defences.

The National Observatory on Climate Change Effects (ONERC) has recommended that managed realignment initiatives need careful planning to account for impacts of climate change on the coasts, such as saltwater intrusion, sea level rise, erosion etc. (ONERC 2007). It seems, however, that in the aftermath of the catastrophic Xynthia storm in 2010, managed realignment has received greater attention according to Philippe Boët[17] (from the National Research Institute of Science and Technology for

---

[16] The latest developments and documentation is found at: http://www.developpement-durable. gouv.fr/Strategie-nationale-de-gestion.html.

[17] Towards depolderisation in the Gironde estuary, http://www.irstea.fr/en/our-editions/info-medias/towards-depolderisation-gironde-estuary#margo.

Environment and Agriculture at Bordeaux). Compulsory land purchase of areas at high risk of flooding is a mechanism used by public bodies in France often after major storm impacts. For example, the *Conservatoire du Littoral* purchased Mortagne-sur-Gironde and Île Nouvelle, two agricultural areas, after they were flooded due to accidental breaching of flood defences during a storm in 1999.

References

Armstrong, S. (2013). *What have we achieved, and what can we learn from elsewhere in Europe?* Presentation at the ABPmer Habitat Creation Conference, 20 November 2013, London.

Bawedin, V. (2000). *Les Bas-Champs de Cayeux/Mer (Somme): enjeux et conséquences d'une éventuelle dépoldérisation.* Master dissertation, Université de Picardie Jules Verne, 151 p.

Bawedin, V. (2004). La dépoldérisation, composante d'une gestion intégrée des espaces littoraux? Prospective sur le littoral picard et analyse À la lumière de quelques expériences: Baie des Veys (Normandie), Aber de Crozon (Bretagne), Tollesbury (Essex) et Freiston shore (Lincolnshire). *Cahiers Nantais, 6,* 11–20.

Beauchard, O., Jacobs, S., Cox, T. J. S., Maris, T., Vrebos, D., Van Braeckel, A., & Meire, P. (2011). A new technique for tidal habitat restoration: Evaluation of its hydrological potentials. *Ecological Engineering, 37,* 1849–58.

Brouwer, R., & van Ek, R. (2004). Integrated ecological, economic and social impact assessment of alternative flood control policies in the Netherlands. *Ecological Economics, 50,* 1–21.

Committee on Climate Change. (2013). *Managing the land in a changing climate.* Chapter 5: Regulating services—Coastal habitats (pp. 92–107). http://www.theccc.org.uk/publication/managing-the-land-in-a-changing-climate/. Accessed 1 Dec 2013.

Cox, T., Maris, T., De Vleeschauwer, P., De Mulder, T., Soetaert, K., & Meire, P. (2006). Flood control areas as an opportunity to restore estuarine habitat. *Ecological Engineering, 28,* 55–63.

De Beukelaer-Dossche, M., & Decleyre, D. (Eds.). (2013). Bergenmeersen, construction of a flood control area with controlled reduced tide as part of the sigma plan. www.vliz.be/imisdocs/publications/250657.pdf. Accessed 18 Feb 2014.

Defra (Department for Environment, Food and Rural Affairs). (2005). *Making space for water—Taking forward a new government strategy for flood and coastal erosion risk management in England.* First Government response to the autumn 2004 consultation exercise. London.

Defra (Department for Environment, Food and Rural Affairs). (2013). *National adaptation programme: Making the country resilient to a changing climate* (p. 182). The Stationary Office, Norwich.

Defra Flood Management Division. (2005). *Coastal squeeze implications for flood management, the requirements of the European birds and habitats directives.* Defra Policy Guidance. https://www.gov.uk/government/uploads/system/uploads/attachment_data/file/181444/coastalsqueeze.pdf. Accessed 3 Feb 2014.

Eertman, R. H. M., Kornman, B. A., Stikvoort, E., & Verbeek, H. (2002). Restoration of the sieperda tidal marsh in the Scheldt estuary, the Netherlands. *Restoration Ecology, 10,* 438–449.

Goeldner-Gianella, L. (2007). Perceptions and attitudes towards de-polderisation in Europe: A comparison of five opinion surveys in France and in the UK. *Journal of Coastal Research, 23*(5), 1218–1230.

Hillen, R., & Roelse, P. (1995). Dynamic preservation of the coastline in the Netherlands. *Journal of Coastal Conservation, 1,* 17–28.

Jacobs, S., Beauchard, O., Struyf, E., Cox, T., Maris, T., & Meire, P. (2009). Restoration of tidal freshwater vegetation using controlled reduced tide (CRT) along the Schelde Estuary (Belgium). *Estuarine, Coastal and Shelf Science, 85*(3), 368–376.

Lawton, J. H., Brotherton, P. N. M., Brown, V. K., Elphick, C., Fitter, A. H., Forshaw, J., Haddow, R. W., Hilborne, S., Leafe, R. N., Mace, G. M., Southgate, M. P., Sutherland, W. A., Tew, T. E., Varley, J., Wynne, G. R. (2010). *Making space for nature: A review of England's wildlife sites and ecological network*. Report submitted to Defra. http://archive.defra.gov.uk/environment/biodiversity/documents/201009space-for-nature.pdf. Accessed 15 Jan 2014.

Meire, P., Ysebaert, T., Van Damme, S., Van den Bergh, E., Maris, T., & Struyf, E. (2005). The Scheldt estuary, a description of a changing ecosystem. *Hydrobiologia, 540,* 1–11.

Ministère de l'Écologie, du Développement durable et de l'Énergie. (2012). Stratégie nationale de gestion intégrée du trait de côte. Vers la relocalisation des activités et des biens. Paris: French Government. http://www.developpement-durable.gouv.fr/IMG/pdf/12004_Strategie-gestion-trait-de-cote-2012_DEF_18–06-12_light.pdf.

ONERC (Observatoire National sur les Effets du Réchauffement Climatique). (2007). Stratégie nationale d'adaptation au réchauffement climatique. La Documentation française (97 p.). http://www.developpement-durable.gouv.fr/IMG/pdf/Strategie_Nationale_2–17_Mo-2–2.pdf.

Patterson, J. (2007). *Coastal Erosion Planning & Response Act (CEPRA)*. Report to the 80th Texas Legislature, February 2007. Texas General Land Office. http://www.glo.texas.gov/what-we-do/caring-for-the-coast/_publications/cepra-report-2007.pdf.

Raynie, R. C., & Visser, J. M. (2002). *CWPPRA adaptive management review final report*. Prepared for the CWPPRA Planning and Evaluation Subcommittee, Technical Committee, and Task Force. http://lacoast.gov/reports/project/3891519~1.pdf.

Rijkswaterstaat, Deltares. (2013). Eco-engineering in the Netherlands: Soft interventions with a solid impact, 44 p.

Rijkswaterstaat, Ecoshape and Deltares. (2013). Eco-engineering in the Netherlands. Soft interventions with a solid impact. Brochure. http://www.deltares.nl/xmlpages/TXP/files?p_file_id=23102. Accessed 3 Feb 2014.

Rupp-Armstrong, S., & Nicholls, R. J. (2007). Coastal and estuarine retreat: A comparison of the application of managed realignment in England and Germany. *Journal of Coastal Research, 23*(6), 1418–1430.

SOGREAH. (2011). Etude de Faisabilite—Depolderisation partielle et eventuelle des Bas-Champs du Vimeu—La recherché d'un avenir sur un territoire perenne. Phase 1: Etat des lieux—diagnostic du territoire. Chapitre 8: Approche géographie sociale—perceptions.

Temmerman, S., Govers, G., Wartel, S., & Meire, P. (2004). Modelling estuarine variations in tidal marsh sedimentation: Response to changing sea level and suspended sediment concentrations. *Marine Geology, 212,* 1–19.

Teuchies, J., Beauchard, O., Jacobs, S., & Meire, P. (2012). Evolution of sediment metal concentrations in a tidal marsh restoration project. *Science of the Total Environment, 419*(1), 187–195.

Van Rompaey, M., & Decleyre, D. (2013). Public support. In M. De Beukelaer-Dossche & D. Decleyre (Eds.), *Bergenmeersen, construction of a flood control area with controlled reduced tide as part of the sigma plan*. Brussels: Agency for Nature and Forest. www.vliz.be/imisdocs/publications/250657.pdf. Accessed 18 Feb 2014.

Chapter 5
Considerations on Managed Realignment in The Netherlands

Joost Stronkhorst and Jan Mulder

Abstract The Netherlands is a small and densely populated country located in the deltas of the rivers Rhine, Meuse, Ems and Scheldt and bordering the North Sea. The low-lying polders behind the dunes and dikes are vulnerable to flooding by rivers and the sea. The dunes, beaches, saltmarshes and intertidal flats are highly valued as natural habitats and recreation areas, but above of all, essential for flood protection. For that reason, sustainable management of these areas is vital and much effort is spent by different authorities concerning with managing the coast. A range of managed options are used along different sectors of the Dutch coast, including the realignment of the shoreline, either landward or seaward, to create space for coastal habitat development and a natural buffer zone for flood protection through dissipation of wave energy. This chapter illustrates the approaches of coastal management applied along the Dutch coast, including an recent experience with managed realignment.

5.1 Introduction

The Netherlands is a small and densely populated country located in the deltas of the rivers Rhine, Meuse, Ems and Scheldt and bordering the North Sea. The Dutch coast consists of three distinct sections (Fig. 5.1): the SW Delta region with islands and enclosed estuaries, the Holland coast with an almost uninterrupted line of dunes and the Wadden Sea region with intertidal flats behind barrier islands and saltmarshes along the mainland. The low-lying polders behind the dunes and dikes are vulnerable to flooding by rivers and the sea. The dunes, beaches, saltmarshes and intertidal flats are highly valued as natural habitats and recreation areas, but above of all, essential for flood protection. For that reason, sustainable management of these areas is vital.

J. Stronkhorst (✉) · J. Mulder
Deltares, PO Box 177, 2600 MH Delft, The Netherlands
e-mail: joost.stronkhorst@deltares.nl

J. Mulder
Water Engineering and Management, University of Twente,
P.O. Box 217, 7500 AE Enschede, The Netherlands
e-mail: jan.mulder@deltares.nl

L. S. Esteves, *Managed Realignment: A Viable Long-Term Coastal Management Strategy?*, 61
SpringerBriefs in Environmental Science, DOI 10.1007/978-94-017-9029-1_5,
© Springer Science+Business Media Dordrecht 2014

Fig. 5.1 Along the Dutch coast, managed realignments through breach of flood defences have been implemented at few locations, including: De Kerf (*1*), Noard-Fryslan Butendyk (*2*), Sieperdaschor (*3*), Breebaart (*4*) and Kroon's polders (*5*). More recently, seaward shoreline realignment projects have been realised along the Holland coast at the Sand Motor (**a**), Hondsbossche zeewering (**b**), Walcheren (**c**) and Zeeuws-Vlaanderen (**d**)

Much effort is spent by different authorities concerning with managing the coast. The national agency Rijkswaterstaat is responsible for shoreline management and flood safety policy, the regional water boards have to provide maintenance of sea defences and nature conservation organisations safeguard wildlife and habitats. One in a suite of management options is managed realignment, i.e. the realignment of the shoreline, either landward or seaward, to create space for coastal habitat development (e.g. saltmarshes) and a natural buffer zone for flood protection through dissipation of wave energy. Figure 5.1 indicates the locations where such projects have been implemented.

Along the Dutch shores, there is little application of managed realignment compared to, for instance, East England. Reasons for this are related to common practices and paradigms rooted in centuries-old heritage of water management (e.g. Van Koningsveld et al. 2008) based on: (i) dike construction and reinforcements and shortening of the coastline for flood safety, (ii) reclamation of land from the sea for agriculture and (iii) 'hold-the-line' (using both hard and soft engineering) to maintain the intrinsic value of dunes and beaches. This chapter will illustrate the approaches of coastal management applied along the Dutch coast, including a brief overview of recent experience with managed realignment.

5.2   Coastline Shortening, Land Reclamation and Hold-the-Line

Along the Dutch coast, dike construction and dune reinforcements have been common practice for centuries to improve flood safety of the low-lying country and to reclaim land that had been lost by successive storm surges. Since the 13th century an estimated 5,700 km^2 of land have been lost to the sea due to many floods caused by storm surges, human-induced land subsidence, coastal erosion and sea-level rise. However, over the centuries and with great effort, land was reclaimed through drainage of lakes or meres in the Holland region and land reclamation of saltmarshes in the Wadden and SW Delta regions. This has created very productive farmland (polders), with a total surface area nearly equal to what was lost (van de Ven 1993).

Historically, driven by the need for safety and land space, 'natural realignments' caused by storm surges have been counteracted. During the last century, this approach has resulted in a systematic shortening of the coastline. A major advantage is that the need for maintenance and reinforcement is restricted to a limited stretch of the coast. In this respect, there are two major projects of dike construction that have played a key role (Bosch and van der Ham 1998). Firstly, the construction of the closure dam Afsluitdijk (1932) created a freshwater lake out of the Zuiderzee reducing the coastline by approximately 600 km. Secondly, in the aftermath of the storm surge in February 1953, which caused a major flood disaster in the SW Delta region, the Delta Project (1958–1997) was initiated. The Delta Project involved the construction of closure dams in all tidal inlets of the area, except the Western Scheldt, shortening the total length of the coastline by 700 km. Consequently, in the last century the coastline of the Netherlands has reduced from 2,000 to 650 km (excluding the coastlines of the Wadden Sea barrier islands).

In 1990, the Dutch government adopted a national policy called dynamic preservation of the coastline (Hillen and Roelse 1995), which in practice aims to 'hold-the-line' as shoreline retreat inland of its 1990 position is not to be allowed (Van Koningsveld and Mulder 2004). The long-term development of the sandy coastal system of the Netherlands is characterized by a structural sediment deficit. Therefore, since 1990 coastline management is achieved by replenishments using sand sources largely available from the deeper North Sea.

In 2000 the policy was extended with the aim of sustainably preserving the sandy coast by keeping pace with sea-level rise. A system-approach was adopted that considers the sand volume of the whole coastal foundation including dunes, beaches and foreshores to a depth of 20 m below mean sea-level and on a timescale of decades to centuries (Mulder et al. 2011). Annually 12 million m^3 of sand are used for foreshore and beach dune nourishments to keep pace with the present sea-level rise of 2 mm/year. This approach is generally regarded by stakeholders as very successful. For the Holland coast in particular, this has changed the sand balance from negative (erosion) to slightly positive (accretion) and consequently resulted in a slight seaward shift of the coastline (Giardino et al. 2014).

Fig. 5.2 (**left**) Beach nourishments and dike-dune reinforcements as part of the *Weak Links* project have improved both flood safety and spatial quality at Nolle-Westduin near Vlissingen, SW Delta. (**right**) The *Sand Motor* project (during construction phase in 2011) involves a large sand nourishment along the eroding coast south of The Hague (photos courtesy of Rijkswaterstaat)

In recent years, innovative coastal defences have been developed that further improve flood safety, beach quality and dune habitats, and imply a shift towards seaward realignments of the coastline. An example is the project *Weak Links* (2008–2015) implemented at eight locations along the coast of Holland and SW Delta, involving a seaward expansion of the coastline by 50–100 m using a combination of hard and soft engineering (Fig. 5.2) in nearly all cases.

An even more substantial seaward expansion of 0.5–1 km was created in 2011 at the eroding coast south of The Hague by a single, large sand nourishment (Fig. 5.2) of 20 million m^3 (Stive et al. 2013). In the next few decades natural processes will distribute the sand gradually to the beach and dunes benefiting this whole coastal stretch thereby strengthen the natural coastal defence. This so-called *Sand Motor* is based on the principles of *building-with-nature* and *eco-engineering* (De Vriend and Van Koningsveld 2012; Rijkswaterstaat, Ecoshape and Deltares 2013) and aims to combine an improved coastal defence and opportunities to create space for nature development and recreation.

Approaches such as the *Sand Motor* are not widely recognised as managed realignment as they involve realignment of the shoreline seawards. However, through shoreline realignment, these approaches create space for a more dynamic response to waves and tides providing opportunities for environmental, social and economic benefits (e.g. enhanced biodiversity, areas for recreation and improved natural coastal defence). Therefore, the *Sand Motor* fits well the broad definition of managed realignment suggested in Chap. 2. Examples of managed landward realignment of the shoreline in The Netherlands are described in the next section.

5.3 Dutch Examples of Landward Realignment of the Shoreline

Given the long tradition of holding the line, the idea of managed realignment as most commonly perceived (i.e. landward realignment) is not accepted lightly in the Netherlands. The issue encounters strong public and political opposition. An example is the planned breach of defences at the Hedwige polder—a 300 ha polder at the Belgium-Netherlands border near Antwerp—as part of a Belgian-Dutch nature-compensation program[1] for dredging activities in the Western Scheldt estuary. After strong opposition in particular from the (farmers) population in the region, in 2012 and after nearly 10 years of political debate, the Dutch government finally decided upon de-poldering, which is scheduled to take place in 2016–2019.

Yet, the potential benefits of managed realignment (i.e. the creation of inter-tidal habitats in response to rising sea levels, thus providing effective wave energy attenuation and contributing to flood safety and reduction of maintenance efforts of the flood defences), have been recognized also in the Netherlands (e.g. Saeijs et al. 2003; ComCoast 2007; Temmerman et al. 2013). The government programme *Room for the River* has formally adopted realignment of river dikes as one of the useful measures to tackle the combined effects of high river discharges and a rising sea level. The programme includes several examples where different managed realignment approaches will be used for flood protection along the rivers in the Netherlands.[2]

Along the coast, over the last two decades, managed realignment has been applied at a few locations as pilot projects. The cases have been initiated by nature conservation organisations and developed in close collaboration with Water Boards, among others, to meet flood safety issues. Implementation of coastal managed realignment projects in the Netherlands can be distinguished based on the types of coastal environments that are being restored:

- in sandy environments along the North Sea coast, projects focus on restoration of natural dune dynamics and biodiversity improvements without hampering flood safety (Löffler et al. 2013) and
- in silty environments along the estuaries and Wadden Sea, projects deal with restoration of saltmarsh dynamics, habitats and biodiversity.

Figure 5.1 shows the location of specific examples of coastal managed realignment pilots in the Netherlands, including:

- De Kerf (Fig. 5.3a), in the calcareous-poor dunes at the Holland coast, between Schoorl and Bergen aan Zee. In 1997, aimed to increase biodiversity and dune dynamics, the seaward dune ridge was breached to allow seawater flowing into the dunes at high tide and the calcareous beach sand being blown landward. More than ten years later, the breach has naturally closed.
- Breebaart (Fig. 5.3b) is a small polder (reclaimed in 1979) near the Dollard estuary. About ten years after embankment the area was purchased to be transformed in a nature conservation site. A connection with the estuary was created in 2001

---

[1] This cross-border project is part of the Sigma Plan, which is the strategy to manage flood risk and nature conservation in the Scheldt estuary and its tributaries (see Sect. 4.2). Information about the Hedwige-Prosper project can be found at: http://www.sigmaplan.be/en/projectareas/hedwige-prosper-project.

[2] Information about the *Room for the River* programme, including the all the nine types of measures that have been formally adopted examples of where they will be implemented can be found at: http://www.ruimtevoorderivier.nl/meta-navigatie/english/publications/.

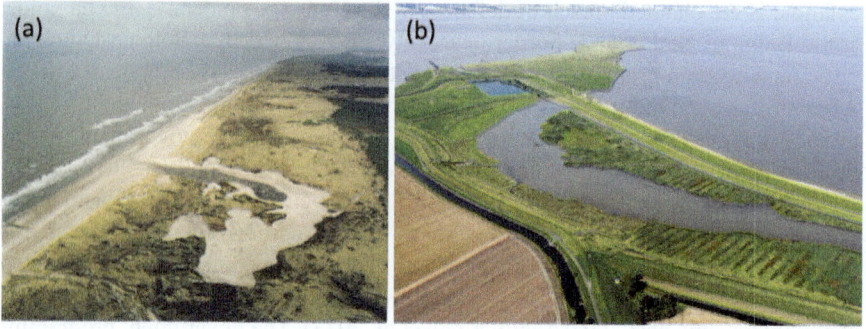

Fig. 5.3 Two examples of small-scale managed landward shoreline realignments in the Netherlands. (**a**) At De Kerf (near Schoorl, North Holland) the seaward dune ridge was artificially breached to increase biodiversity and coastal dynamics. (**b**) At the Breebaart polder (near the Dollard estuary in rural North Netherlands) a culvert and a fish passage restored tidal flow into the embanked area resulting in an increase in biodiversity (photos courtesy of Rijkswaterstaat)

using a culvert and a fish passage. This has improved the fish population, salt-marsh vegetation and increased the number of waders.[3]

- Noard-Fryslan Butendyks, summer polders along the Wadden Sea coast of North Friesland. Summer polders are reclaimed saltmarshes now used to herd cattle during the summer. These polders are bordered by dykes, higher in the landward side (to protect the mainland from flooding) and lower seaward. De-poldering by breaching the seaward dike started at Noarderleech in 2001 and Bildt Pollen in 2009. Results so far indicate enhanced diversity of flora and fauna.[4] A total of 4,000 ha of saltmarsh is planned to be restored at these sites.

- Sieperdaschor[5] is a small polder reclaimed in 1966 along the Western Scheldt estuary at the Belgian-Dutch border, adjacent to the large saltmarsh area named Land van Saeftinge. A storm in February 1990 breached the seaward dike (therefore it is not an actual 'managed' realignment). As the polder was of little value, the breach was left open for saltmarsh development. The initial sedimentation rate after breaching reached 4–8 cm/year (Kornman 2000).

- Kroon's polders,[6] at the Wadden Sea barrier island Vlieland. These few small polders were created in the early 20th century by developing a series of dikes using branches and reed screens to trap sand. However, the polders were of little agricultural use due to a high groundwater table and became a nature conservation area. In 1996, the Wadden dike was breached to restore tidal flow into the polders.

---

[3] The website with information about the reserve provides information about the biodiversity (in Dutch): http://www.avifaunagroningen.nl/index.php/gebieden/569-polder-breebaart.

[4] The website with information about the reserve provides photos and information about the biodiversity (in Dutch): http://www.itfryskegea.nl/Natuurgebied/Noard-Fryslan-Butendyks/.

[5] Information about the site can be found (in Dutch) at: http://www.deltabirding.nl/gebied.asp?g=33.

[6] Find out more about the history and biodiversity of the site at: http://www.ecomare.nl/en/encyclopedia/regions/wadden-sea-region/dutch-wadden-region/vlieland/vlie-nature/kroons-polders/.

5.4   Conclusions

The removal of sea defences to create space for wetland development is to be a critical issue for the low-lying delta of the Netherlands. Managed realignment leading to a landward shift of the shoreline goes against a long tradition of dike construction, land reclamation, shortening of the coastline and the national policy of 'hold-the-line'. Using the concept of *building-with-nature* and *eco-engineering*, over the last decade the focus is shifting towards a gradual seaward expansion of the shoreline. This approach stimulates wetland development in an offshore direction, as illustrated by the *Sand Motor* project. Projects creating a deliberate seaward shift of the shoreline position are not widely recognised as managed realignment. However, they can be considered a method of managed realignment under the wider definition adopted in this book. Over the last two decades, albeit at a small scale, managed realignment through breach of defences or controlled tidal restoration has been applied along the Dutch coast mainly with the aim of improving biodiversity and restoring natural dynamics along sandy (i.e. dunes along the North Sea coast) and silty shorelines (i.e. along estuaries and Wadden Sea).

References

Bosch, A., & van der Ham, W. (1998). Twee eeuwen Rijkswaterstaat 1798–1998. Europese Bibliotheek Zaltbommel, 343 p.

ComCoast. (2007). Work package 3 Civil engineering aspects—ComCoast flood risk management schemes. Final Report. Accessed 18 Feb 2014. http://ec.europa.eu/ourcoast/download.cfm?fileID=773.

De Vriend, H. J., & Van Koningsveld, M. (2012). *Building with nature: Thinking, acting and interacting differently.* EcoShape, Building with Nature. Dordrecht.

Giardino, A., Santinelli, G., & Vuik, V. (2014). Coastal state indicators to assess the morphological development of the Holland coast due to natural and anthropogenic pressure factors. *Ocean & Coastal Management, 87,* 93–101.

Hillen, R., & Roelse, P. (1995). Dynamic preservation of the coastline in the Netherlands. *Journal of Coastal Conservation, 1,* 17–28.

van Koningsveld, M., & Mulder, J. P. M. (2004). Sustainable coastal policy developments in the Netherlands. A systematic approach revealed. *Journal of Coastal Research, 20*(2), 375.

van Koningsveld, M., Mulder, J. P. M., Stive, M. J. F., van der Valk, L., & van der Weck, A. W. (2008). Living with sea level rise; a case study of the Netherlands. *Journal of Coastal Research, 24*(2), 367.

Kornman, B. A. (2000). Sieperdaschor: 10 years of morphological research. Rijkswaterstaat document RIKZ/OS/2000/80x, 14 p. (in dutch).

Löffler, M., van der Spek, A. J. F., & van Gelder-Maas, C. (2013). *Options for dynamic coastal management—A guide for managers* (Report 1207724). Deltares: Deltares, Bureau Landwijzer, Rijkswaterstaat Centre for Water Management.

Mulder, J. P. M., Hommes, S., & Horstman, E. M. (2011). Implementation of coastal erosion management in the Netherlands. *Ocean and Coastal Management, 54,* 888–897. doi:10.1016/j.ocecoaman.2011.06.009.

Rijkswaterstaat, Ecoshape and Deltares. (2013). Eco-engineering in the Netherlands. Soft interventions with a solid impact. Brochure. http://www.deltares.nl/xmlpages/TXP/files?p_file_id=23102. Accessed 3 Feb 2014.

Saeijs, H., Smits, T., Overmars, W., & Willems, D. (2003). *Changing estuaries, changing views.* Nijmegen: Erasmus University. http://hdl.handle.net/1765/1850.

Stive M., de Schipper M., Luijendijk A., Ranasinghe R., van Thiel de Vries J., Aarninkhof S., van Gelder-Maas C., de Vries S., Henriquez M., Marx, S. (2013). The sand engine: A solution for vulnerable Deltas in the 21st century? *Proceedings Coastal Dynamics, 2013,* 1537–1545. Available from http://www.coastaldynamics2013.fr/pdf_files/148_Stive_marcel.pdf. Accessed 18 Feb 2014.

Temmerman, S., Meire, P., Bouma, T. J., Herman, P. M., Ysebaert, T., & De Vriend, H. J. (2013). Ecosystem-based coastal defence in the face of global change. *Nature, 504,* 79–83.

van de Ven, G. P. (Ed.). (1993). *Man-made lowlands a history of water management and land reclamation in the Netherlands.* The Netherlands: Utrecht Matrijs—Ill. ISBN 90-5345-030-0.

Chapter 6
The National Trust Approach to Coastal Change and Adaptive Management

Phil Dyke and Tony Flux

Abstract The National Trust is a UK-based non-government charity that aims to protect open spaces and historic places. The Trust's properties are held in perpetuity for the enjoyment of the public and at present they include 1,187 km of coastlines (almost 10 % of the shoreline) in England, Wales and Northern Ireland. The scale of the Trust's coastal ownership presents serious management challenges associated with sea-level rise, in particular an increase in coastal erosion and flooding. Principles of sustainability and integrated coastal managed underpins the work of the Trust, which includes, wherever possible, to allow a natural roll-back of the shoreline and planning to move out of the risk areas. Our challenge is to effectively communicate the long-term benefits of this approach. This chapter describes the principles and approaches adopted by the Trust and provide an example of removal of coastal defences at Brownsea Island, southern England.

6.1 Introduction

The National Trust[1] is a charity totally independent of Government that aims to protect open spaces and historic places. The Trust's properties are held in perpetuity for the enjoyment of the public. The very first donation to the Trust in 1895 was a 2 ha of gorse-covered cliff top at Dinas Oleu overlooking Cardigan Bay in Wales. Today the Trust cares for 1,187 km of coastlines (almost 10 % of the shoreline) in England, Wales and Northern Ireland, including dunes, saltmarshes, soft cliffs, hard cliffs, as well as villages, infrastructure, harbours etc.

The scale of the Trust's coastal ownership presents serious management challenges associated with sea-level rise, in particular an increase in coastal erosion

---

[1] More about the National Trust is found at: http://www.nationaltrust.org.uk/what-we-do/.

P. Dyke (✉)
National Trust, Heelis, Swindon, Wiltshire, SN2 2NA, UK
e-mail: Phil.Dyke@nationaltrust.org.uk

T. Flux
National Trust, Filcombe Farm, Morecombelake, West Dorset, DT6 6EP, UK
e-mail: tony.flux@nationaltrust.org.uk

L. S. Esteves, *Managed Realignment: A Viable Long-Term Coastal Management Strategy?*, 69
SpringerBriefs in Environmental Science, DOI 10.1007/978-94-017-9029-1_6,
© Springer Science+Business Media Dordrecht 2014

and flooding. To deal with these challenges in a sustainable manner, the Trust has adopted the following coastal management principles:

- The coast is dynamic and changing and we will work with natural processes wherever possible, taking a long-term view and flexible approach to coastal management, enabling and promoting adaptability in response to sea level rise.
- Interference in natural coastal processes will only be supported for reasons of overriding benefit to society. We will only favour development at the coast which takes full account of coastal change and environmental/cultural considerations (including landscape/seascape).
- Valued natural and cultural coastal heritage features will be conserved as far as is practicable. We accept some valued features will be lost and we will substitute if appropriate (natural assets) or properly record where loss is inevitable (cultural values).
- We aspire to deliver Integrated Coastal Zone Management (ICZM) in our sites and will work with others to achieve this.
- The Trust promotes public access and recreational opportunity at the coast to help realise health and well-being benefits and increase people's understanding of the coast and marine environment.
- The Trust will consider acquisition where it is the best option to support these principles.

6.2 National Trust Understanding and Planning for Coastal Change

Forecasts suggest with increasing confidence that climate change will lead to continued sea-level rise and increased storminess, accelerating the scale and pace of coastal change. To help plan for this uncertain future, the Trust commissioned a study looking at how our coastline is likely to change over the next 100 years. The results lead to the production of the *Coastal Risk Assessment* 1, a gazetteer examining flood, erosion and accretion responses to a 1 m rise in sea level at the Trust's coastal sites. Subsequently, the Coastal Risk Assessment Phase 2 (CRA2) completed a more detailed analysis of these impacts at each of our 500 coastal sites. The main essence of the CRA2 reports is presented to the public in the publication *Shifting Shores* (National Trust 2005), an unprecedented policy document that sets out the Trust's position on working with a changing coast. Companion documents looking in more detail at coastal change in Wales, South West England (National Trust 2007) and key sites in Northern Ireland (National Trust 2008) were also published.

CRA2 results paint an interesting and perhaps unexpected picture of the future. Around the Trust 70 'hotspot' locations (Fig. 6.1) were identified, where a complex mix of issues may pose a challenge to management, including threats to infrastructure, habitats, historic structures, communities and third-party interests. To support the management of these complex situations, a decision-making framework

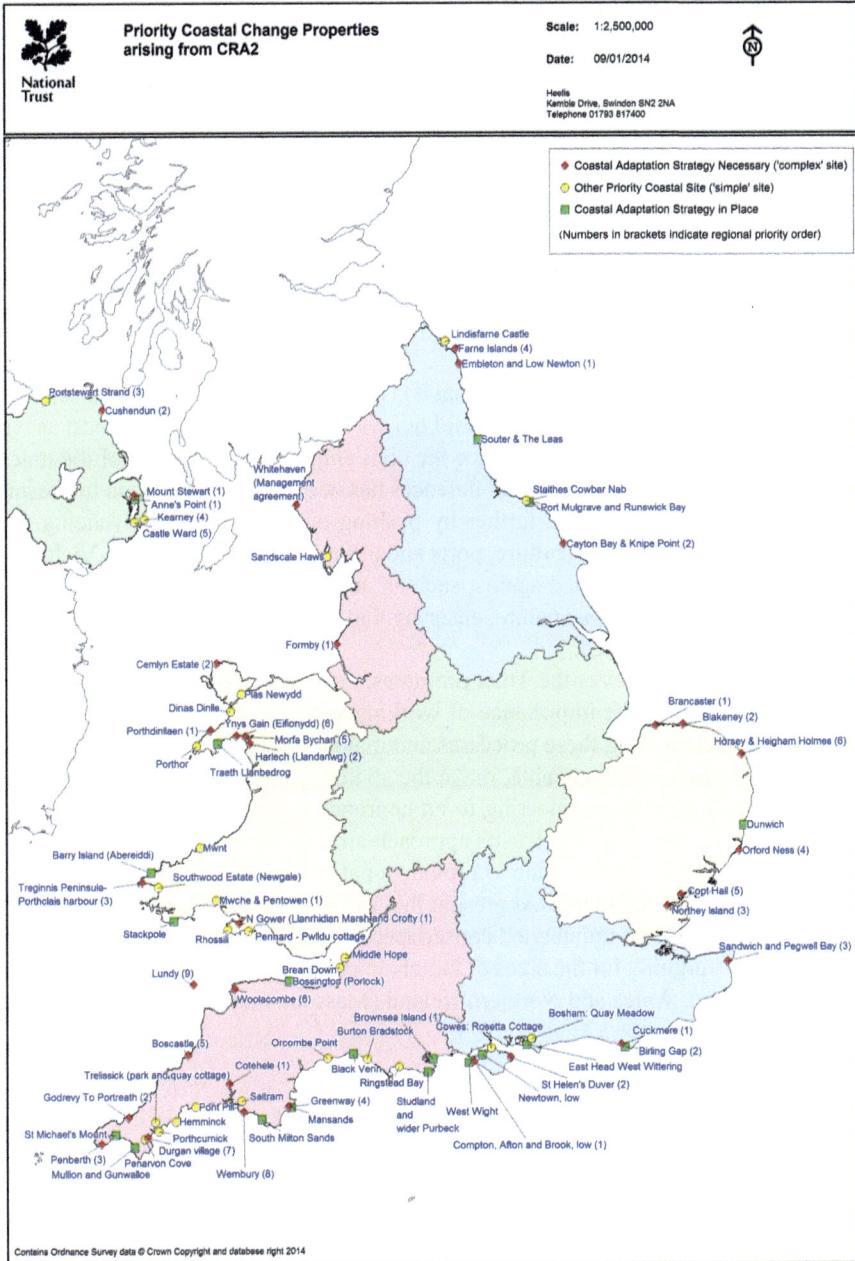

Fig. 6.1 Location of the National Trust coastal management 'hotspots' identified in the CRA2

in now in place to develop *Coastal Adaptation Strategies* (CAS) that, over time will cover all hotspot locations. The actions identified in each CAS are intended to feed back into the Trust's core business planning mechanisms and inform/be informed by relevant public policies (e.g. Shoreline Management Plans; Flood and Coastal Erosion Risk Management Strategies[2] and the yet untested Coastal Change Management Areas).[3]

6.3 Hard Choices and Coastal Squeeze—The Advocacy Message

The big message in *Shifting Shores* is that it is unrealistic, in a time of rising sea levels, to think that we can continue to build our way out of trouble on the coast as we have for the past 150 years. Ever since the civil engineering advances of the nineteenth century, building hard coastal defences has been the usual reaction to coastal change. Indeed we have gone further by pushing out into the sea, advancing the line, reclaiming land for agriculture, ports and urban settlements. Our pre-Victorian forebears may have cautioned against such an audacious approach, counselling us to work *with* and not *against* nature, ensuring that coastal infrastructure was placed beyond the grasp of the sea.

Through *Shifting Shores* the Trust promotes a discussion at a national, regional and local level about the importance of working with natural coastal processes. It is through understanding these processes and making space for change at the coast that we can, wherever practicable, make the switch from a shoreline management response based on civil engineering to an approach based on natural process. Instead of pouring concrete, the Trust's approach aims to enable newly eroded material to be retained at the shoreline to provide a natural sea defence in the form of a beach, sand dune or saltmarsh. At present this valuable sediment is too often being washed out to sea as an unintended consequence of engineered coastal defences.

The Trust is arguing for the areas of largely undeveloped coast, about 70 % of the coast of England, Wales and Northern Ireland (Masselink and Russell 2013), to be

---

[2] Under the requirement set by the Flood and Water Management Act 2010, the Environment Agency has developed a strategy for flood and coastal erosion risk management in England. The strategy establishes Lead Local Flood Authorities (usually the role is taken by local authorities) and requires that they develop local flood risk management strategies taking an integrated and co-ordinated approach. Similarly, a National Strategy was developed for Wales. The national strategies for England and Wales can be accessed, respectively, from the following links:

http://www.official-documents.gov.uk/document/other/9780108510366/9780108510366.pdf
http://wales.gov.uk/docs/desh/publications/111114floodingstrategyen.pdf.

[3] The Planning Policy Statement 25 (Development and Coastal Change) published in 2010 requires Local Planning Authorities to identify Coastal Change Management Area, "where rates of shoreline change are significant over the next 100 years, taking account of climate change". Shorelines where the SMP indicate the policies of hold-the-line or advance the line (maintain existing defences or build new defences) are excluded.

able to evolve in a natural manner. Shoreline features, such as sandy dunes, should be free to naturally move inland as sea level rises and, subject to sediment supply, maintain a natural and dynamic form of sea defence. If constrained by a man-made sea wall on one side and a rising sea-level on the other, coastal features become squeezed, with sediment transported offshore and often lost. Coastal squeeze will leave behind a steepened coastal profile and an undermined sea wall vulnerable to collapse. The Trust is not alone in highlighting this concern. The Marine Climate Change Impacts Partnership (MCCIP 2008) highlighted that, over the past 100 years, intertidal profiles have steepened in almost half of England's and in a quarter of Wales' shoreline, particularly on coasts protected by hard engineering structures.

Future shorelines dominated by failed sea defences and denuded of sand will not be attractive places to visit. From a coastal tourism perspective, it is also attractive to have naturally functioning coastlines for people to enjoy—an asset to support the often fragile economy of rural coastal communities. Following careful consideration, at a number of properties, the Trust decided to remove failing or counterproductive sea defences and to allow natural coastal processes to resume.

6.4   Making Time and Space for Change

In the future there may be a place for sea-defences but in the Trust we are clear that these structures will only be appropriate as a mechanism to enable us to buy time so we can develop long-term and sustainable approaches to manage our future coast. In some places this may involve managed re-alignment through seawall breaching on reclaimed land or unpicking failed, failing and counterproductive sea defences. More generally it will be about allowing a natural roll-back of the shoreline and planning to move out of the risk areas wherever we can. Working with natural processes and have adaptability as guiding principles inevitably means that some coastal management decisions will be difficult and on occasion controversial. Our challenge is to effectively communicate the long-term benefits of this approach.

A key issue for the Trust and wider society is the need to be able to think for the long-term if we are to deal effectively with a changing coast. Thinking in 20, 50 and 100-year time-frames needs to become the norm, to replace our current practice of thinking in short 5-year planning and political cycles. Our *Coastal Management Principles* and *Coastal Adaptation Strategies* are enabling staff at the Trust's coastal sites to put this thinking into practice—challenging ourselves to think beyond the short-term. The *Shifting Shores* documents illustrate how the Trust is adopting novel approaches to the conservation of buildings, the management of visitor infrastructure, creating opportunities for wildlife and ensuring we understand the significance of archaeological features as they are exposed.

To help individuals and organisations to face the realities of coastal change, national governments need to continue investing in coastal change monitoring and research. When financial resources are tight, knowing where to get the best return from coastal defences investments if of paramount importance. Such information

Fig. 6.2 Aerial view of Brownsea Island showing the extent of the shoreline restoration project. The tidal inlet of Poole Harbour (which opens to the east-southeast) is seen at the *top right (image courtesy of the National Trust)*

can only be gained by understanding how our coasts are evolving. A comprehensive network of coastal monitoring observatories is in place in England and Wales (e.g. the Channel Coast Observatory[4]). In Northern Ireland there is still a need to develop such monitoring initiative and a parallel process of shoreline planning—an approach that the National Trust is actively seeking to bring about through working with others. The Trust believes that we stand the best chance of making the right decisions about the future management of our coast by: utilising the best available scientific information; embracing an adaptive approach; harnessing creativity and innovation; and communicating effectively with all neighbours, stakeholders and partners.

6.5 Brownsea Island Shoreline Restoration Project

To illustrate the Trust's approach in practice, the actions taken to enable a more naturally functioning shoreline at Brownsea Island are described here. This example highlights the challenges associated with unpicking failed sea defences, which is one of the many ways of facilitating coastal realignment.

---

[4] The Channel Coastal Observatory is the data management centre for the Regional Coastal Monitoring Programmes, it compiles and provides a range of real-time and field data that can be accessed online from: www.channelcoast.org.

Fig. 6.3 (**a**) Failing defences in Brownsea Island (**b**) were removed in 2011 and (**c**) by early 2013 the shoreline showed an appearance similar to the time before sea defences were built

Brownsea Island (Fig. 6.2) has been in the ownership of the National Trust since 1962 and is one of the most popular coastal sites in the Trust's care. The Island has an area of 202 ha and it is located within Poole Harbour (one of the largest natural harbours in the world), in Dorset, South West of England (Fig. 6.1). During the 1970s, various attempts were made to impede erosion along 2.4 km of the south and west shores, including the use of: 105 m of steel pilings, 2,600 wooden piles and numerous stone-filled gabions. The defences were erected at a time when the traditional approach to coastal erosion was to defend by whatever means available. In 2008 all these coastal defences were in a failing condition (Fig. 6.3a). The defences were unsightly and the wooden pilings in particular represented a potential hazard to navigation if they broke away and floated into the navigation channels.

The Brownsea Island Management Approach adopted in 2009 recognised the exceptional wildlife and heritage value of the Island and sought to offer safe access to all permitted areas for visitors, including most of the foreshore. Failing and unsightly sea defences conflicted with this aim and spoiled the ambience of the southern foreshore. Ensuring safety for the public and navigation also helped justified the removal of the failed defences. Taking into consideration the Trust's coastal management principles, a project aiming to restore a more naturally functioning shoreline started in 2009.

The decision to remove the failed sea defences was taken based on the Coastal Adaptation Strategy for Brownsea[5] and three additional assessments conducted in 2009 (Failing Defences Condition Survey; Consulting Engineers Brief; and Brownsea Island Sea Level Rise Study). The conclusions of the three studies indicated that the accelerated erosion projected to occur due to the removal of coastal defences would be acceptable. The findings indicated that the eroded sediment would be beneficial to build beach levels and contribute to the natural evolution of the foreshore profile.

Staff, volunteers and the public were concerned to hear that there was no intention to replace the failing defences. Time and effort were invested in meetings and workshops[6] to engage all stakeholders in discussing the logic behind the idea that an

---

[5] Further information can be found at: http://www.climatebuffer.eu/downloads/ldv_visit_brownsea_island_shoreline-120926.pdf.

[6] An educational video was produced with support from the Department for Environment, Food and Rural Affairs (Defra) to help explaining the project during meetings. The video is available from: http://www.youtube.com/watch?v=ifDGs_eAIZ8.

undefended shoreline was the most sustainable management option for that section of the Brownsea shoreline.

6.5.1 Shoreline Planning, Consents and Designations

The development of the Brownsea shoreline restoration project took place within the context of a number of shoreline planning, consenting and designation regimes:

Shoreline Planning The Shoreline Management Plan for the area (Royal Haskoning 2010) indicates 'no active intervention' policy for this section of coast for the next 100 years. It was clearly beneficial to have a broad alignment between the public policy and the Trust's intentions for the management of this shoreline.

Consents According to the legislation in place in 2010, it was necessary to establish whether or not the project required a Food and Environmental Protection Act Licence and Coastal Protection Act Consent. However, the marine licensing system in England changed in the 6th April 2011 in accordance with provisions made in the Marine and Coastal Access Act 2009. Therefore, a number of communications with the Marine Management Organisation officers were required to clarify the licensing issue before it was established that an exemption clause *did* apply due to the minimalist nature of the intervention.

The proposed demolition works required planning permission and pre-planning advice was sought from the Purbeck District Council in September 2010, with permissions granted four months later. A number of ecological and archaeological conditions needed to be complied with during the project. The Trust also sought agreement with Poole Harbour Commissioners, the harbour operating authority, that the work should take place. The Harbour Commissioners have responsibility for ensuring navigational safety and were supportive, in particular, that the wooden piles were removed.

Designations Brownsea Island forms part of the Poole Harbour Site of Special Scientific Interest (SSSI). SSSI are areas of conservation protected by national legislation. The south shore of the island is specifically designated for important littoral sediments and the associated flora and fauna. Until recently, environmental assessments reported this shoreline as being in 'unfavourable condition', partly due to the old coast defences interfering with natural sediment transport processes.

The National Trust applied for a Higher Level Stewardship (HLS) grant to secured sufficient funds to implement the complete programme of works. HLS grants are managed by Natural England and funds projects capable of delivering significant environmental benefits in priority areas. A re-appraisal of the environmental conditions in the Poole Harbour SSSI concluded that the area of the project (sub-unit 60) is now in 'unfavourable–recovering' condition[7].

---

[7] SSSI conditions countrywide is assessed by Natural England and results are available online. For Poole Harbour check: http://www.sssi.naturalengland.org.uk/special/sssi/unit_details.cfm?situnt_id=1030590.

In January 2011 thousands of wooden pilings and gabions started to be removed (Fig. 6.3). Work halted in March 2011 to prevent interference with the bird breeding season and re-commenced in November 2011 when all the steel pilings were removed. As part of the programme, most of the material was recycled and the foreshore was graded to something akin to a natural profile. Within three months, the gradient of the sandstone cliffs were beginning to readjust and beach levels were noticeably improved. By early 2013, the south shore had returned to its pre-intervention appearance (Fig. 6.3c). Monitoring shows that the sediment cover is still improving and leading to enhancement of the amenity value.

6.6 Brownsea and Beyond: Looking to the Future

The Brownsea shoreline restoration project illustrates that managed realignment can take a number of different forms. At Brownsea removal of the failed sea defences was 'managed' and cliffs and foreshore are now realigning. Coastal management, as the coast itself, is constantly evolving as innovative solutions are implemented to respond more sustainably and efficiently to climate change and sea-level rise. Pressure to 'defend' urban frontages will increase as existing engineering works break down or become obsolete. The costs of replacing these defences will rise, requiring an ever more critical appraisal through cost-benefit analysis. As money is channelled to protect areas with the most valuable economic assets, it will be increasingly difficult to secure funding for sea defences along rural and less developed coastlines.

Managed realignment and other approaches for re-naturalising coastlines are likely to become the most plausible option on less developed coasts. These solutions are not necessarily cheap to implement but they offer the chance for transformation. Re-establishment of a naturally functioning shoreline free ourselves from the 'sea defence cycle'—construct, fail and reconstruct. In the long-run we must adapt and move out of the risk areas wherever we conceivably can rather than storing up the problems for future generations to deal with.

It is heartening to see some signs that public policy in the UK is moving beyond a narrowly focused reliance on hard engineering solutions for shoreline management. One example of moving beyond the 'defend or do nothing' stalemate is the approach articulated in the publication *Working with natural processes to manage flood and coastal erosion risk* (Environment Agency 2010). Although, and as welcoming as this thinking is, it seems there is a significant disconnect in the 'line of sight' between innovative high level strategy and actions on the ground. The latter often being undertaken in response to storm events when the necessity emerges for public bodies to be seen to apply yet another short term fix rather than engaging communities in more sustainable solutions based on roll-back, realignment and restoring the natural functioning of our shorelines.

At many National Trust 'coastal change hotspots', the journey of managed realignment is still to begin. Guided by the Trust's coastal management principles, adaptive approaches to the coastline will continue to be promoted and adopted in coastlines under our care.

References

Environment Agency. (2010). Working with natural processes to manage flood and coastal erosion risk. http://cdn.environment-agency.gov.uk/geho0310bsfi-e-e.pdf. Accessed 10 Jan 2014.

National Trust. (2005). Shifting shores: Living with a changing coastline. http://www.nationaltrust.org.uk/document-1355766981061/. Accessed 10 Jan 2014.

National Trust. (2007). Shifting shores in the South West: Living with a changing coastline. http://www.nationaltrust.org.uk/document-1355766981103/. Accessed 10 Jan 2014.

National Trust. (2008). Shifting shores: Living with a changing coastline in Northern Ireland. http://www.nationaltrust.org.uk/document-1355766980464/. Accessed 10 Jan 2014.

Masselink, G. & Russell, P. (2013). Impacts of climate change on coastal erosion. Marine Climate Change Impacts Partnership: Science review 2013, 16 p. http://www.mccip.org.uk/media/13234/2013arc_backingpapers_9_ce.pdf. Accessed 10 Jan 2014.

MCCIP (Marine Climate Change Impacts Partnership). (2008). Marine Climate Change Impacts Annual Report Card 2007–2008. http://www.mccip.org.uk/media/3007/arc2007.pdf.

Royal Haskoning. (2010). Poole & Christchurch bays shoreline management plan. http://www.twobays.net/.

Chapter 7
Managed Retreat in Maui, Hawaii

Thorne Abbott

Abstract Beach and marine-based activities are of great importance to Maui's economy, culture and lifestyle. Hard engineering structures hinder access to the beach, impound sand reservoirs and contribute to beach scour, leading to narrow beaches that can ultimately be completely lost. Beach loss equates to a financial loss in the long-term. Maui County is taking forward managed retreat initiatives to facilitate the maintenance of a wide beach able to support tourism and to reduce the risk from extreme storm events to people and property. Therefore, managed retreat in Maui is a policy-decision underpinned by both economic and environmental reasons. This chapter describes current policy approaches and provides examples of managed retreat measures in Maui.

7.1 The Beach: an Economic Asset

Maui County is taking forward managed retreat initiatives in two forms: relocation after major storm impact and through the implementation of shoreline setbacks based on historic erosion rates. Beach and marine-based activities are of great importance to Maui's economy, culture and lifestyle (Fletcher et al. 2003). Hard engineering structures hinder access to the beach, impound sand reservoirs and contribute to beach scour, leading to narrow beaches that can ultimately be completely lost. Beach loss equates to a financial loss in the long-term. Managed retreat is implemented to facilitate the maintenance of a wide beach able to support tourism and to reduce the risk from extreme storm events to people and property. Therefore, managed retreat in Maui is a policy-decision underpinned by both economic and environmental reasons.

Maui's beaches are golden yellow consisting of moderately coarse coralline sands. Nearshore waters are crystal clear and blue oceans stretch to neighboring islands where the water's surface is frequently breached by whales within the *Hawaiian Islands Humpback Whale National Marine Sanctuary* during the winter season. It is to everyone's advantage to preserve this unique coastal experience, and to profit from its naturalness.

T. Abbott (✉)
Coastal Planners, LLC, 3993 Maalaea Bay Place, Wailuku, HI, USA

L. S. Esteves, *Managed Realignment: A Viable Long-Term Coastal Management Strategy?*,   79
SpringerBriefs in Environmental Science, DOI 10.1007/978-94-017-9029-1_7,
© Springer Science+Business Media Dordrecht 2014

Currently, Maui is undertaking a study to support the development of guidelines and protocols to promote coastal redevelopment and reconstruction after a major coastal storm, hurricane or tsunami. The study is assessing relocation options for private and public infrastructure and the preferred development patterns after an acute shore-changing event. A decision matrix based on coastal typology, development patterns and environmental qualities is being proposed to help expedite the reconstruction permitting process and reduce negative impacts on coastal amenities. Recognising the particularities of the range of coastal typologies found in the County's three islands[1] is an integral part of the process. The matrix would be vetted with local, state and federal regulators, and determinations of 'appropriate' redevelopment patterns evaluated through community-based workshops, including remote and underserved communities. Results of the study are anticipated in January 2015.

7.2   Implementation of Setback Lines

Maui adopted shoreline setbacks in 2003 based on site-specific erosion rates[2] (Fletcher et al. 2003). Thereafter, seven resorts and a number of residential homes redeveloped their property by demolishing structures along the shoreline and locating new development inland of the shoreline setback area. When a private property is slated for redevelopment, the government approving agency can seek exactions of the developer by requiring new structures to be located inland of the defined shoreline setback to protect public health and safety. The setback is calculated based on the annual rate of erosion[3] that is likely to occur over the expected lifespan of the building; which is considered as 50 years in Maui, and 70 years in Kauai (Abbott 2013). This enables the developer to assure potential property buyers that their investment is relatively safe from damage that would be caused by beach erosion and flooding.

Most of Maui's turist accomodation were built in the 1970s and 1980s and are now outdated in relation to current standards. Acute erosion events in 2003, 2005 and 2007 caused damage to resorts on Maui's west shore resulting in a decrease in visitors and tourism revenue. The hoteliers realised that without a quality sandy beach experience, their occupancy and room rates would quickly decline. It was a financial decision that led resorts to relocate out of harm's way as they refurbished and reconstructed oceanfront properties.

---

[1] The County of Maui consists of the islands of Maui, Lānai, and Molokai (part of Molokai is within Kalawao County).

[2] The rules of shoreline setbacks in Maui are explained at: http://www.co.maui.hi.us/index. aspx?NID&=&865.

[3] Annual rates of erosion were calculated by researchers from the Coastal Geology Group at the University of Hawaii and results can be viewed at: http://www.soest.hawaii.edu/coasts/erosion/maui/.

Fig. 7.1 Oceanfront buildings (highlighted in the *left* and *top right*) of the ANdAZ Hyatt, former Wailea Renaissance, were demolished and new development is seen under construction in a position landward of the demolished buildings their position is highlighted (*bottom right*). Images sources: (*left*) Wailea Renaissance; (*top right*) orthorectified aerial photography dated of 1997, Coastal Geology Group, University of Hawaii; (*bottom right*) Google Earth, imagery date: 4 Mar 2013

The Wailea Renaissance (now ANdAZ Hyatt), along Maui's south shore, removed an underground sewer trunk line and demolished three buildings; and new development has been built landward of the setback area (Fig. 7.1). A seaside concrete walkway is being replaced with recycled materials that are portable and movable and allow easy repositioning following erosion. A new emergency vehicle access path, increased public beach parking and other public amenities are also part of the redevelopment.

Along Halama Street, a new development involved demolishing two older homes, constructing two luxury oceanfront homes (landward of the shoreline setback), and removing a 45 m long, 2.5 m high series of geo-textile bags that were installed on the properties seaward edge. The area was restored with native vegetation to allow for a more natural grade and permit the previously impounded sand resources to respond naturally to coastal dynamics.

At Charlie Young's beach, a popular tourist location in Kihei, the public beach was substantially narrowed by the presence of *naupaka*[4] shrubs grown by homeowners as a privacy buffer at the seaward edge of their properties. The hedge

---

[4] *Scaevola taccada*, a native Hawaiian plant

Fig. 7.2 At Charlie Young's beach the beach width and dunes have been restored after a campaign resulted in removal of shrubs grown by beachfront property owners over the public beach (photos: Tara M. Owens, County of Maui Planning Department)

hindered lateral access along the beach during high tide. A public education and outreach campaign resulted in the removal of these hedges and restoration of the site to a natural sand dune (Fig. 7.2). Community volunteering work contributed to improve dune restoration by erecting sand fences and by installing dune walkovers to focus foot traffic and reduce dune blowouts and facilitate access for those with disabilities. These efforts have doubled the beach width enhancing public amenity value and restoring the natural adaptive capacity to respond to erosion events.

7.3 Concluding Remarks

A long, wide, natural beach is more attractive to tourists and residents than one that is narrow, intermittent and riddled with concrete seawalls. Managed retreat emphasises asset management, natural capital and ongoing accrued interest from beach resources, rather than cashing in on the asset's immediate value. Planning to avoid coastal development in areas prone to erosion and flooding based on a site's historical rates of shoreline retreat provides a rational nexus for relocating development out of harm's way. Managed retreat reduces the risk to property damage and the public's health and safety, protecting resources held in trust for the public and future generations.

References

Abbott, T. (2013). Shifting shorelines and political winds—The complexities of implementing the simple idea of shoreline setbacks for oceanfront developments in Maui, Hawaii. *Ocean & Coastal Management, 73,* 13–21.

Fletcher, C., Rooney, J., Barbee, M., Lim, S. C., & Richmond, B. (2003). Mapping shoreline change using digital orthophotogrammetry on Maui, Hawaii. *Journal of Coastal Research, SI 38,* 106–124.

Chapter 8
Managed Realignment in the UK: The Role of the Environment Agency

Karen Thomas

Abstract The Environment Agency is responsible for the strategic overview for flood and coastal management for England. With partner organisations, the Environment Agency has delivered over 50 managed realignment projects in the UK in the past three decades. In result, by the end of 2014, over 900 ha of intertidal habitat will have been created through managed realignment. Future strategic plans include realignment of 10% of the country's coastline by 2030. This chapter describes the drivers for the implementation of managed realignment as a mechanism for delivering flood risk and environmental policies, and the opportunities and challenges that exist. A description of lessons learned so far and the role of emerging ecosystems services and biodiversity offsetting approaches to support managed realignment strategies is also provided.

8.1 Introduction

Discussion about managed realignment started in the UK in the late 1980s. At the time practitioners were questioning the need for ongoing maintenance of flood defences in some locations around the coast due to rising costs. In particular, consideration was given to locations where there were relatively self-contained rural areas of land, rising in elevation from the floodplain with no people, property or infrastructure. At the same time coastal squeeze had been identified as an issue and saltmarsh habitat was in decline so there were calls to recreate intertidal habitat areas.

By the mid-late 1990s the selection of early managed realignment sites was based on the conditions of the flood defences and challenges faced to maintain them (e.g. exposure to high wave energy or undercutting by tidal currents) and the willingness of landowners to the new approach. Intertidal habitat and new flood storage areas were created as a result.

The Environment Agency (EA) is responsible for the strategic overview for flood and coastal management for England (EA 2007). Partner organisations and the EA have successfully delivered managed realignment at about 50 UK locations

Karen Thomas (✉)
Environment Agency, Iceni House,
Cobham Road, Ipswich, IP3 9JD, UK
e-mail: Karen.thomas@environment-agency.gov.uk

L. S. Esteves, *Managed Realignment: A Viable Long-Term Coastal Management Strategy?*,   83
SpringerBriefs in Environmental Science, DOI 10.1007/978-94-017-9029-1_8,
© Springer Science+Business Media Dordrecht 2014

Fig. 8.1 (**left**) Managed realignment at Medmerry (West Sussex, southern England). The beach ridge was breached in November 2013 and the target is to create around 320 ha of intertidal habitat once the project is completed. (**right**) The Steart Coastal Management Project under construction (Somerset, southwest England) is the largest managed realignment site in the UK to date (over 400 ha); breach is to occur in 2014

in the past three decades, resulting in the creation of 400 ha of intertidal habitat. By the end of 2014, a further 500 ha of new intertidal habitats will be created with the completion of the second phase of Medmerry and the new Steart Peninsula site, the two largest managed realignment sites in the UK to date (Fig. 8.1).

This chapter will describe the EA's role in implementing managed realignment as a mechanism for delivering broader flood risk and environmental policy agendas and the opportunities and challenges that exist. The text identifies legislative and policy drivers and discusses the use of emerging ecosystems services and biodiversity off-setting approaches to support managed realignment. The importance of engaging communities, landowners and partners is considered, especially in the light of re-sults from a survey about lessons learnt to date from managed realignment practices (Thomas 2013) conducted for the EA programme entitled *Better Ways of Working*.

8.2 Legal and Policy Drivers for Managed Realignment

Defra's biodiversity policy sets as a minimum standard that all flood risk manage-ment works must be environmentally acceptable (Defra 2011). The EA is required to seek opportunities for environmental enhancement when selecting flood and coastal defence options both at the strategic and project level. Defra's biodiversity outcome measures are reflected in the EA's corporate strategy[1] as performance tar-gets. To achieve these targets, a strategic view is needed, both to deliver the most cost-effective managed realignment and to maximise opportunities to create habi-tats from the start of the planning process. Biodiversity is at the centre of both legal and policy drivers leading to the implementation of managed realignment. The main drivers are described next.

---

[1] The EA's corporate reports, including updates, can be accessed from: http://www.environment-agency.gov.uk/aboutus/149594.aspx.

8.2.1   EU Habitats and Birds Directives

On behalf of the UK Government the EA is responsible for ensuring that all possible actions are taken to avoid further net loss of protected intertidal habitats. As part of this legislative duty, the EA needs to maintain Natura 2000 sites and habitats that support protected bird species and assemblages. Therefore, any flood-risk management scheme which unavoidably affects Natura 2000 sites requires creation of habitat as a compensatory measure. In some places, the only way to protect people and properties from flood risk is through new or improved flood defences. Where upgrading of existing flood defences or construction of new ones are proposed within designated Natura 2000 sites damage or loss of intertidal habitats must be compensated.

The legal requirement of protecting Natura 2000 sites also takes into consideration long-term processes leading to habitat loss and degradation (e.g. due to coastal squeeze) and both historical and future losses count to estimate the extent of compensatory habitat that needs to be created. A recent review on the implementation of the Habitats and Birds Directives (Defra 2012a) indicates that the ecosystems approach can and should play a greater role in assisting long-term strategic decisions about mitigation and compensation measures.

8.2.2   EU Water Framework Directive (WFD)

The WFD requires consideration of the potential impacts of retaining flood defences around our coast and estuaries on water quality and the features that support it. The WFD requires improvements to the ecological status of transitional (estuarine) and coastal water bodies and a need to retain or increase vital intertidal habitats. Where flood defences are constraining natural estuary functions, the implementation of WFD may actively support managed realignment as a remedy against coastal squeeze and to improve the ecological status of transitional and coastal water bodies. In the UK, WFD policy is developed through River Basin Management Plans (RBMPs), which are the highest tier of strategic plans promoting delivery of sustainable water management including flood and coastal management. There is also guidance to develop WFD assessments as part of Shoreline Management Plans (SMPs) (e.g. East Anglia Coastal Group 2010). RBMPs give a steer on managed realignment as an option, especially where broader water management outcomes align with SMP. The project *Managed Realignment: Moving Towards Water Framework Objectives*[2] (funded by the EU LIFE programme) used existing evidence to provide guidance for practitioners on how managed realignment can lead to the improvement of ecological quality in estuaries meeting the standards set by the WFD.

---

[2] Project reports are available from: http://www.environment-agency.gov.uk/homeandleisure/floods/123710.aspx.

8.2.3   UK Government Policy on Flood Risk Management and Habitat Creation

Defra encourages a strategic (e.g. estuary-wide) approach to deliver intertidal habitat re-creation, including compensation for sea-level rise. As the EA manages several thousand kilometres of flood defences around the coast of England, coastal squeeze is an issue that must be addressed to meet environmental obligations. A strategic approach helps to conserve sites subject to long-term habitat loss and avoid delays in implementing justified flood management works. RBMPs and SMPs are best placed to provide the necessary business case, strategic framework and rationale for undertaking habitat creation for compensation and environmental enhancement. The creation of larger areas of new habitat, rather than several isolated smaller areas, is encouraged as a more sustainable approach both ecologically and economically.

Regional Habitat Creation Programmes (RHCP) proactively seeks willing landowners to develop managed realignment projects ahead of capital projects and ongoing maintenance needs of EA's flood defences. The EA then acquire sites, as and when they become available. This strategic approach helps reduce land purchase costs to the taxpayer and allows planning, in collaboration with landowners and partners, to maximise project outcomes. The RHCP approach was first implemented in the Anglian region in 2003 and has since been replicated across EA's regions. These regional programmes have a system of recording, reporting and accounting for habitat created and provide a national view of habitat gains and losses. The recording system allows identifying whether there is a pipeline of projects in place to balance current and future coastal change. The RHCP system is auditable and transparent to clarify the links with the statutory drivers and outcomes.

The strategic approach described here has been successful. However, delivering managed realignment is challenging. Finding willing landowners is difficult. Gaining support from communities and local politicians can be very hard. Often managed realignment is perceived as 'giving up land to the sea' (see Chap. 2), especially where landowning families have been involved in local flood defence management in the past. New legislation and innovative ways to assess the value of the natural environment (e.g. ecosystem services) are creating opportunities to improve and adapt our approach to managed realignment.

8.2.4   Emerging Policy

The report *Making Space for Nature: A review of England's Wildlife Sites and Ecological Network* (Lawton et al. 2010) sets out a number of recommendations that will shape future government's directions about nature conservation. The report suggests that to face climate change impacts (e.g. sea-level rise and increased storminess) natural defences will be more important than ever. The need for dynamic coastal environments is highlighted and, to promote dynamic coastlines, investments in 'soft engineering' solutions are required. The importance of eco-system services is emphasised, including the role of saltmarshes in natural flood regulation, carbon sequestration and nutrients cycling.

One of the biggest challenges in delivering managed realignment is the engage-ment of willing landowners. Creating new funding streams to incentivise land-owners to embrace managed realignment might be an option to address this issue. Lawton et al. (2010) recommend that the government should promote the creation of new markets and stimulate payment for ecosystem services to ensure these are taken into account in decisions that affect the management and use of the natural environment.

> The urgent and logical next step is to develop markets that enable these values to be realised for services such as water quality, flood risk management, climate regulation and other benefits.If we take into account the potential values of a broad range of ecosystem services, the benefits of establishing and managing a coherent and resilient ecological network could, in many situations, outweigh the costs many times over. There is an urgent need to develop market mechanisms through which landowners can realise the value of the ecosystem ser-vices that their land provides to society (Lawton et al. 2010, p. 83).

Payment for ecosystem services is now under consideration by Defra as a mecha-nism for funding habitat management (Dunn 2011; Scott Wilson 2011). There has been an increase in interest in recent years in the creation of new, small-scale mar-kets to fund nature conservation management (Scott Wilson 2011). There is already evidence that people are prepared to pay a premium for products that bring benefits for wildlife and other environmental values, including more sustainable ways of farm and land management (Dunn 2011). Understanding the links between biodi-versity and a wider range of ecosystem services is rapidly improving and we are increasingly able to place values on such services (Ecosystem Markets Task Force 2013; Defra 2013).

Defra are currently piloting ecosystem services approaches, for example, to sup-port the development of flood risk management policies in a community-led estuary management plan in the Deben Estuary in Suffolk. The Deben Estuary Partnership, the EA and the Suffolk Coast and Heaths Unit are working together in the develop-ment of the Deben Estuary Plan.[3] In demonstrating the benefits that ecosystems have to offer and establishing which flood risk management policies will enhance ecosystems, communities and partners are better able to identify the most sustain-able options. By identifying the benefits of flood risk management approaches, including managed realignment, it is possible to map potential beneficiaries who in turn may be interested in investing in the delivery of the plan. Such approach supports the Sustainability Appraisal process, which is an essential element of the plan. The experience gained in the development of the Deben Estuary Plan and the extensive work already conducted on the Humber Estuary will underpin how the Government and the EA will use ecosystem services in the future.

---

[3] Information about the partnership and the Deben Estuary Plan can be found at: http://www.suf-folkcoastandheaths.org/projects-and-partnerships/deben-estuary-partnership/.

Biodiversity offsetting has been suggested as a way of enhancing nature and creating wider benefits through planning (Lawton et al. 2010; Ecosystem Markets Task Force 2013; Defra 2013). Biodiversity offsets are designed to compensate for residual and unavoidable harm to existing wildlife sites caused by development. It is fundamental to biodiversity offsets the production of measurable outcomes based on losses and gains being quantified in the same way (Defra 2012b). This mechanism aims to ensure that, when a development causes detrimental impact on nature (and this damage cannot be avoided), new, bigger or better nature sites will be created. Biodiversity offsetting could be a mechanism used to support managed realignment in estuarine and coastal areas to create new coastal wetlands (ICE Maritime Expert Panel 2013).

In 2012, Defra started a two-year pilot programme for biodiversity offsetting[4] at six locations (managed by Natural England, Defra and local councils): Devon, Doncaster, Essex, Greater Norwich, Nottinghamshire, Warwickshire, Coventry and Solihull. The lessons learnt from these pilots will inform UK Government policy. In September 2013, Defra started an open consultation asking about the role biodiversity offsetting could play in mitigating and compensating for environmental damage as a result of development in England. The responses to the consultation are currently being analysed and the outcome will soon be published.[5]

Biodiversity offsetting has the potential to be more broadly and strategically used. Developers could invest in intertidal restoration projects in a similar way that some private companies invest in 'green' projects as part of their wider business portfolio. In particular, carbon and nutrient sequestration opportunities that saltmarsh re-creation offer would be directly comparable with other carbon offsetting initiatives that are now common-place in corporate business.

Ecosystem services approaches are already being employed in managed realignment projects and at the strategic level in development of flood risk management plans. Biodiversity offsetting is an emerging mechanism that could provide further support to managed realignment in the future. The current EA corporate strategy will be reviewed in 2015 and any emerging policy will be reflected in the 2016–2020 strategy.

8.3   What Have We Learnt from Existing Managed Realignment Projects?

Many managed realignment projects are currently in place or underway. However, within the existing policy and legislative framework, the ease and speed of managed realignment delivery can be challenging. Local opposition has prevented the imple-

---

[4] Further details about biodiversity offsetting in the UK can be found at: https://www.gov.uk/biodiversity-offsetting.

[5] The consultation outcomes will be available from: https://www.gov.uk/government/consultations/biodiversity-offsetting-in-england.

Fig. 8.2 (**left**) Managed realignment at Abbotts Hall (Blackwater estuary, Essex) and (**right**) Paull Holme Strays, Humber estuary (photos courtesy of the Environment Agency)

mentation of managed realignment even in areas where realignment has already been implemented, such in Essex and on the Humber estuary (Fig. 8.2).

In 2013, the EA has decided to review existing experiences with the objective to inform and improve works related to flood and coastal risk management. As part of the *Better Ways of Working Programme* the review aimed to establish a snapshot of the current thinking among practitioners about managed realignment projects that have been delivered to date. A questionnaire was developed and distributed to key practitioners (within the EA and externally) with experience in managed realignment projects asking their views on lessons learnt and opportunities to take forward new ways of working. A relatively small number of practitioners have actually managed realignment projects and their experience and knowledge is valuable and needed to be documented.

The EA is using the outputs of the questionnaire survey to inform practitioners and encourage further efficiencies and innovative approaches to managed realignment projects in the future. To build in-house capacity and expertise, the EA also takes into consideration evidence produced by other studies, such as the independent survey conducted by Bournemouth University (Esteves and Thomas 2014). The EA survey covered four broad aspects relevant to the implementation of managed realignment projects and a summary of the outcomes in each of these aspects is presented below.

8.3.1   Engagement and Communication

Since the late 1990s, due to changing legislation, the primary objective for managed realignment has shifted from flood risk management to an option to create compensatory habitat. Currently there is a general view that emphasising managed realignment as a multifunctional approach might make it more appealing to the public and stakeholders and help attract funds from a wider range of sources. By engaging with local communities to identify how they are likely to benefit from future managed realignment, a greater sense of ownership might be created leading to increased uptake of managed realignment in general and better acceptance of projects near their homes and businesses.

The reasons for delivering managed realignment are varied and there is no single justification that may be appropriate to all projects. However, it is important that practitioners are able to communicate a clear and consistent message about why managed realignment is needed here and now. It is important to consider how stakeholders and partners perceive managed realignment so a communication strategy can be developed to address common misunderstandings and false expectations. Public engagement and uptake will be even more important to support the emerging role of managed realignment in delivering Water Framework Directive and climate change adaptation outcomes (Committee on Climate Change 2013).

Tools and techniques available to practitioners for engagement and communication are well advanced. However, having time to engage and reach the relevant groups has been identified as challenging. Opportunities to enhance engagement should be sought where possible, particularly with regard to the messages we give about managed realignment. Sometimes even policy names may lead to misconceptions. For example, managed retreat was often used interchangeably with managed realignment but fell in disuse as it was considered to give a negative connotation to the policy (see Chap. 2). *Intertidal wetland restoration* is emerging as a more favourable policy name to replace managed realignment.

At many locations, public perception has been challenging, sometimes delaying or preventing the implementation of local managed realignment projects (e.g. the second breach Deveraux farm in Essex and Donna Nook on the Humber). Local communities sometimes are uncertain about the benefits they might accrue from managed realignment. Public opposition is a problem especially when projects involve loss of access rights to existing footpaths or there is a perception that increased flood risk will occur elsewhere in their area. Public perception is clearly a communication issue that needs to be addressed; even more so as it also affects partnership and funding opportunities.

8.3.2 Statutory Process

The process for obtaining statutory consents and licences required for managed realignment are often perceived as a significant challenge. Since managed realignment projects are still relatively rare compared to the more routine planning applications on the coast, planners and regulators need support at an early stage to increase knowledge and understanding. The publication of the latest SMPs should help to underpin decision-making at planning stages. Whilst the SMP is not a statutory planning document, most Local Authorities have adopted or endorsed SMP policies (including managed realignment) and embedded these policies into core planning documents (e.g. Local Plans and Core Strategies). Consequently, there is policy guidance to support statutory consultees, which should improve the planning process for future managed realignment projects.

Understanding the timescales of statutory processes is key to avoiding delays later in the project. The EA has a role to work closely with statutory partners to streamline the process so that the various planning and marine consents are aligned (e.g. planning permission, marine licences, sustainability or strategic environmental appraisal). Individually these can each take several months so they need to be planned in parallel where possible. Straightforward managed realignment projects with good partnerships and engaged communities can be delivered with 2–3 years (e.g. Abbotts Hall) but more typically take longer if popular footpaths and perceived flood risk issues are involved (e.g. Deveraux farm Essex and Donna Nook Humber respectively). Whilst there will be local differences, the RHCP approach means that knowledge can be shared across England between EA practitioners and partners to ensure a consistent approach is employed.

8.3.3 Technical Expertise

Practitioners in the UK have the knowledge and expertise to develop and create managed realignment projects. Technical expertise seems to be the aspect in which practitioners have gained the most experience and confidence, especially in terms of selecting, engineering, monitoring and modelling managed realignment projects. However, it is important to recognise that only few practitioners have direct experience of managed realignment so sharing tools and techniques is a good way to build and enhance capacity for the future.

The EA will be seeking opportunities to build knowledge and experience through networks and sharing of good practice. The EA intends to compile and centralise relevant information and guidance, which will be then signposted to EA and external practitioners. In addition, sharing data obtained from monitoring is key to developing robust evidence whilst reducing costs. This is already under development with Natural England and can easily be extended to other organisations and partners.

As discussed in Sect. 8.2.4, managed realignment is likely to play an important role in the implementation of biodiversity offsetting opportunities, ecosystems services and the Water Framework Directive. As new objectives and requirements emerge, the development of appropriate technical expertise will be needed.

8.3.4 Funding and Partnership

The EA has specific outcomes for managed realignment sites for which Government funds are available. The EA have successfully worked with landowners and non-Government organisations, most notably the Royal Society for the Protection of Birds (RSPB) and The Wildlife Trusts, to deliver wider environmental and socio-economic outcomes. Experience has shown that working in partnership is the best

Fig. 8.3 Breaching of flood defences at Wallasea Island (Crouch estuary, Essex) was completed on the 4th July 2006. The project was implemented in partnership with the landowner and the RSPB (photos courtesy of the Environment Agency)

way forward. Good examples to date include: Abbotts Hall (Fig. 8.2) involving partnership with the Essex Wildlife Trust; and many projects delivered in partnership with the RSPB, such as Freiston in Lincolnshire, Medmerry (Fig. 8.1) on the southern Kent coast and Wallasea Island (Fig. 8.3) in Essex.

Partnerships allow for shared knowledge to strengthen the potential for success in terms of project delivery and the likelihood of local acceptance. Shared knowledge facilitates the identification and dissemination of the wider benefits managed realignment can deliver and partners contribute to select the appropriate design to maximise chances for the realisation of the intended outcomes. Partnerships generally lead to a more streamlined project delivery and help attracting wider funding streams, supporting the objective to create more sustainable coasts that can adapt to climate change and sea-level rise.

Finding willing landowners is a challenge that leads to the conundrum of strategic *versus* opportunistic habitat creation. Current SMPs and Estuary plans indicate the locations where managed realignment is the most sustainable policy options. However, the willingness of landowners is central to enable the strategic view to be delivery at the local level. managed realignment projects cannot progress without willing landowners. Landowners can choose to maintain flood defences with private funds in locations where flood defences are no longer economic (and therefore no longer maintained by the Government). As a result it may be challenging to deliver managed realignment in the most sustainable locations.

If willing landowners are identified in less suitable locations, the EA may choose to implement managed realignment at these more opportunistic sites due to the need to meet habitat creation targets. However, opportunistic managed realignment schemes do not always deliver the best return for the investment of public money and there is a need to balance opportunity against the benefits that can be realised. Results of a survey conducted within the *Managing Coastal Change Project*[6] indicates that landowners are not necessarily adverse to managed realignment as a policy but they express that greater business incentives are needed to encourage them to 'give up' productive agricultural land for intertidal habitat creation.

The ecosystem services approaches currently being piloted by Defra (see Sect. 8.2.4) identify the type of benefits that can be realised from managed realignment and the potential beneficiaries. Such approaches may provide funding mechanism opportunities for future managed realignment projects. Managed realignment is likely to play an important role in the implementation of the Water Framework Directive, biodiversity offsetting and ecosystems services approaches. Improved understanding of the wider benefits provided by managed realignment in the light of emerging policies is expected to strengthen project outcomes and enhance the appeal to partners, funders and communities.

8.4 Conclusions

The EA is required to deliver intertidal habitat due to legal requirements and to implement Government policy through RHCP. The EA contributes significantly to managed realignment developments in England and, through managed realignment, will have created over 900 ha of intertidal habitat by 2015. Funding and partnership working is fundamental to the delivery of managed realignment at the strategic and local level. Working in partnership is the best way to ensure the delivery of multi-functional coastal wetlands and to gain trust and support from local communities and stakeholders. Emerging ecosystems services and biodiversity offsetting approaches are under consideration and, following policy guidance from Defra, these will be reflected in the EA Corporate plan 2016–2020, which will steer future directions of managed realignment implementation. The EA recognises the importance of managed realignment as an approach for delivering a wide range of environmental, socio-economic and flood risk management outcomes. The EA will continue to work with landowners, communities and partners to promote and develop managed realignment projects with the objective of delivering the greatest benefits to society while meeting statutory environmental requirements.

---

[6] The project was funded by Making Space for Water-Defra 2007 Innovation Fund, which was set up to share £ 1.2 million across UK to non-government projects concerning adaptation to climate change.

References

Committee on Climate Change. (2013). Chapter 5: Regulating services—coastal habitats (pp. 92–107). http://www.theccc.org.uk/wp-content/uploads/2013/07/ASC-2013-Chap5_singles_2.pdf. Accessed 10 Jan 2014.

Defra. (2011). Biodiversity 2020. A strategy for England's wildlife and ecosystem service. https://www.gov.uk/government/uploads/system/uploads/attachment_data/file/69446/pb13583-biodiversity-strategy-2020-111111.pdf. Accessed 16 Jan 2014.

Defra. (2012a). Report of the habitats and wild birds directives implementation review. https://www.gov.uk/government/uploads/system/uploads/attachment_data/file/69513/pb13724-habitats-review-report.pdf. Accessed 14 Jan 2014.

Defra. (2012b). Biodiversity offsetting pilots. Technical paper: The metric for the biodiversity offsetting pilot in England. https://www.gov.uk/government/uploads/system/uploads/attachment_data/file/69531/pb13745-bio-technical-paper.pdf. Accessed 18 Jan 2014.

Defra. (2013). Realising nature's value: The final report of the ecosystem markets task force. Government response. https://www.gov.uk/government/uploads/system/uploads/attachment_data/file/236879/pb13963-government-response-emtf-report.pdf. Accessed 16 Jan 2014.

Dunn, H. (2011). Payments for ecosystem services. Defra evidence and analysis series, paper 4. https://www.gov.uk/government/uploads/system/uploads/attachment_data/file/69329/ecosystem-payment-services-pb13658a.pdf. Accessed 16 Jan 2014.

East Anglia Coastal Group. (2010). The wash shoreline management plan. Appendix K water framework directive assessment. http://www.eacg.org.uk/smp4.asp. Accessed 15 Jan 2014.

Ecosystem Markets Task Force. (2013). Realising nature's value: The final report of the ecosystem markets task force. http://www.defra.gov.uk/ecosystem-markets/files/Ecosystem-Markets-Task-Force-Final-Report-.pdf. Accessed 16 Jan 2014.

Environment Agency. (2007). Coastal strategic overview—implementation plan. Sea flooding and coastal erosion risk management. www.archive.defra.gov.uk/environment/flooding/documents/policy/strategy/coastimp.pdf. Accessed 14 Jan 2014.

Esteves, L. S., & Thomas, K. (2014). Managed realignment in practice in the UK: Results from two independent surveys. *Journal of Coastal Research,* Special Issue 70, 407–413.

ICE Maritime Expert Panel. (2013). The role of coastal engineers in delivering no net loss through biodiversity offsetting—A discussion paper. http://www.ice.org.uk/Information-resources/Document-Library/The-Role-of-Coastal-Engineers-in-Delivering-No-Net. Accessed 18 Jan 2014.

Lawton, J. H., Brotherton, P. N. M., Brown, V. K., Elphick, C., Fitter, A. H., Forshaw, J., Haddow, R. W., Hilborne, S., Leafe, R. N., Mace, G. M., Southgate, M. P., Sutherland, W. J., Tew, T. E., Varley, J., & Wynne, G. R., (2010). Making space for nature: A review of England's wildlife sites and ecological network. Report submitted to Defra. http://archive.defra.gov.uk/environment/biodiversity/documents/201009space-for-nature.pdf. Accessed 15 Jan 2014.

Scott Wilson, H. (2011). Barriers and opportunities to the use of payments for ecosystem services. Final report to defra. http://randd.defra.gov.uk/Document.aspx?Document=PESFinalReport28September2011(FINAL).pdf. Accessed 16 Jan 2014.

Thomas, K. (2013). A review of managed realignment. Sharing and embedding coastal knowledge. Report for the environment agency programme better ways of working, October 2013.

Chapter 9
Factors Influencing the Long-Term Sustainability of Managed Realignment

Nigel Pontee

Abstract Managed realignment can offer a sustainable long-term management option for coasts and estuaries by reducing the pressures on flood defences and compensating for habitat lost due to developments or coastal squeeze. This chapter discusses some of the factors influencing the long-term sustainability of managed realignment taking into consideration the UK context and experience. The sustainability of managed realignment at the local level depends on the availability and suitability of land to deliver set objectives and greatly on stakeholder support. The implementation of more than 50 schemes in the last two decades in the UK indicates that these factors have not hindered the implementation of managed realignment at the national scale. However, managed realignment is not the best option for all coastal areas. It is certain that managed realignment schemes will continue to be built in the UK. However, managed realignment is not a universal panacea, it is not possible to implement this policy everywhere, and schemes are unlikely to be viewed as a success from all perspectives. External funding sources and extensive stakeholder consultation are required to support the scale of managed realignment implementation planned in the future.

9.1 Introduction

Managed realignment is considered here in its most common form, which involves re-introducing tidal regimes to areas of previously reclaimed low-lying land. This can be undertaken by breaching or removing the existing flood defences; or by using structures (e.g. culverts and tidal gates) to create regulated tidal exchange (RTE) schemes. Managed realignment typically creates a range of mudflat, saltmarsh and transitional habitats.

Managed realignment helps reducing flooding and erosion risks by allowing coastal habitats to migrate inland in response to climate change and sea-level rise. Managed realignment can therefore help limit the projected increase in the cost of providing coastal defences in the long-term (Adaptation Sub-Committee 2013) and plays an important role in the management of our coasts (Pontee and Parsons 2012).

N. Pontee (✉)
Global Technology Leader Coastal Planning & Engineering, CH2M HILL, Burderop Park, SN40QD Swindon, UK
e-mail: Nigel.Pontee@ch2m.com

L. S. Esteves, *Managed Realignment: A Viable Long-Term Coastal Management Strategy?*, 95
SpringerBriefs in Environmental Science, DOI 10.1007/978-94-017-9029-1_9,
© Springer Science+Business Media Dordrecht 2014

In the UK, more than 50 managed realignment schemes have been implemented (see list in the Appendix) and an estimated 1,300 ha of new intertidal habitat were created (Adaptation Sub-Committee 2013).

In order to discuss whether managed realignment is sustainable in the long-term, it is necessary to understand the main policy drivers to date and the direct or indirect influencing factors at the strategic and implementation levels. In the UK, two main drivers lead to the implementation of managed realignment:

Provision of compensatory habitats required under European legislation: The Habitats Directive (European Commission 1992) and the Birds Directive (European Commission 2009) have been important drivers for restoring coastal habitats in Europe (see Pontee et al. 2013 for further details). Port expansion projects reclaiming intertidal mudflat habitat, or construction of sea defences on existing saltmarsh, commonly need to provide compensatory habitat, including for indirect losses arising from coastal squeeze. In the UK, compensation measures have commonly been delivered by recreating new habitats within managed realignment schemes.

Strategic flood and coastal erosion management: In England and Wales the central Government plays a key role in setting the policy framework for flood and coastal defence (see Chap. 8). The strategic approach has been implemented through a three level hierarchy of Shoreline Management Plans (SMPs), strategy plans and schemes (e.g. Pontee et al. 2005; Pontee and Parsons 2010). The idealised route for a flood risk management scheme would therefore be: (1) for the SMP to recommend this as a policy, (2) a strategy to confirm the policy and to identify the type of scheme that will deliver the policy, and (3) a scheme to carry out the detailed design, seek necessary approvals (e.g. planning permission etc.) and implement the works. Strategies typically investigate net gains and losses of existing habitats and make recommendations about how and where managed realignment should be implemented. Further details on assessing the gains and losses of estuary habitats can be found in Canning and Pontee (2013). Strategic approaches to the management of coastal habitats for some coasts and estuaries have been promoted by Coastal Habitat Management Plans (CHaMPS, led by Natural England) and specific habitat management or managed realignment studies funded by the Environment Agency or Local Authorities.

The most recent round of SMPs and strategies in England and Wales developed policies for three epochs (0–20, 20–50 and 50–100 years). These time-scales were deliberately chosen in recognition that present day 'hold the line' policies may not be sustainable, but that coastal communities need time to adapt to more sustainable polices such as managed realignment. According to the Adaptation Sub-Committee (2013) the managed realignment policies proposed in the SMPs for the English coast will create approximately 6,200 ha by 2030 and 11,500 ha by 2060.

The two drivers of (i) compensatory habitat provision and (ii) coastal defence management are interlinked. Additionally, when analysing schemes that have been constructed, confusion may arise between the driver (i.e. the primary reason why

Fig. 9.1 Alkborough Flats scheme in the Humber Estuary. This scheme was opened in September 2006 with the principal objectives being to create new intertidal habitat and provide flood storage (photo: www.petersmith.com, courtesy of the Environment Agency)

a scheme is undertaken) and the resulting benefits. Whilst some schemes, such as Welwick[1] (Humber estuary), have been undertaken purely to recreate habitat lost to port development in the estuary, many other schemes have involved both drivers in some way. The following examples illustrate this:

- The Alkborough scheme in the Humber Estuary (Fig. 9.1) was driven by the need to provide compensation for habitat loss directly and indirectly caused by flood defences throughout the estuary, but also by the opportunity to reduce extreme water levels and flood defence costs elsewhere in the estuary (see Wheeler et al. 2008).
- Hesketh Out Marsh West in the Ribble Estuary (Fig. 9.2) was developed to create additional intertidal habitat to compensate for habitat losses in Morecambe Bay, while maintaining flood defence standards in the local area of the Ribble (see Tovey et al. 2009).
- The Steart Scheme in the Parrett Estuary (Fig. 9.3) was undertaken primarily to provide compensatory habitat for losses caused by sea defences in the Bristol Channel and Severn Estuary, but also allowed the provision of improved flood defences to the village of Steart and its access road.

---

[1] Breaching of flood defences at Welwick MR site occurred in 2006 and this 54 ha scheme forms part of the compensation package for port developments at Immingham and Hull.

Fig. 9.2 Hesketh Out Marsh West in the Ribble Estuary during the construction phase. The scheme covers about 168 ha and was opened in 2008 with the principal objective being to create compensatory intertidal habitat (photo courtesy of the RSPB)

Fig. 9.3 Steart Coastal Management Project in the River Parrett is due to be opened in autumn 2014. This scheme has two objectives: to provide compensatory habitat and to improve flood defences for the local community. Note that the breach has not yet occurred at the time of the photo, the water seen within the scheme is freshwater due to a particularly wet winter (photo courtesy of the Environment Agency)

9.2 Factors Influencing the Sustainability of Managed Realignment

Sustainable shoreline management policies have been defined as "those which take account of the relationships with other defences, developments and processes, and which avoid, as far as possible, committing future generations to inflexible and expensive options for defence" (Defra 2006, p. 12). However, in the real world, practical understanding of sustainability typically varies between stakeholders depending on their interest or perspective. This is especially the case for managed realignment schemes since the natural habitats created can perform a number of functions (see Table 1.1). The sustainability of schemes can be considered from any of these perspectives. A recent review stated that the ecosystem services provided by coastal habitats, could equate to a value of £ 680–2,500 per ha (Adaptation Sub-Committee 2013). The key factors influencing the sustainability of managed realignment are identified and discussed below.

9.2.1 Creation of Habitats Within Realignment Sites

Since managed realignment schemes are often used to provide compensatory habitat it is important to consider how the habitat created within schemes compares with natural habitats. The successful creation of intertidal habitats hinges on the creation of appropriate physical conditions within sites. Key physical aspects include a number of hydraulic, hydrological, morphological and sedimentological factors (see Pontee 2003).

Site elevation is one of the most fundamental considerations in the design of intertidal habitats. Elevation influences the frequency and duration of tidal inundation, as well as exposure to wave action, all of which affect the colonisation and development of vegetation (Pontee 2003). Given this, it is perhaps not surprising that inappropriate elevations has been one of the most significant factors in the failure of a number of schemes in the USA in the past (Roberts 1991). In the UK, experience has shown that saltmarsh recreation proceeds best between elevations of mean high water neap (MHWN) tides and mean high water spring (MHWS) tides, which equates to around 450–500 tidal inundations each year. Furthermore, a gradual, 'natural' gradient across the marsh surface provides a range of elevations and tidal inundations, which promotes a more diverse saltmarsh (Toft and Madrell 1995).

Site elevation is a dynamic parameter influenced by sedimentation levels in the site. In estuaries with high suspended sediment concentrations such as the Humber or the Severn, it can be difficult to create sustainable, long-term mudflat habitat within schemes. This is because the high sediment supply coupled with relatively quiescent conditions within the realignment site can lead to high siltation rates, which cause mudflats to accrete vertically and evolve into saltmarsh habitats (Morris 2012; Pontee 2014). Such habitat changes have implications for

Fig. 9.4 Extensive areas of reeds have colonised the inner parts of Poole Harbour. The presence of reeds means that there is a high likelihood of them colonising any managed realignment sites that are created in this area (photo: Nigel Pontee)

invertebrate assemblages and bird usage (ABP Research & Consultancy Ltd 1998; Atkinson et al. 2001; Nottage and Robinson 2005).

Mazik et al. (2010) have concluded that realignment sites may only be a short-term solution to the loss of intertidal mudflats, owing to high natural accretion rates within the sites resulting in a change from mudflat to saltmarsh. On the Humber, high siltation rates have been observed in the many realignment sites particularly in the early years following opening. For example, at Alkborough the first year of monitoring recorded rates of up to 0.58 m. Similarly, at Paull Holme Strays rates of up to 0.17 m were measured over a three-month period. The design issues associated with high siltation rates are described further in Pontee (2014).

In relation to the use of managed realignment to re-create saltmarsh habitats, Mossman et al. (2012) found that the resulting plant communities were significantly different in their composition from natural marshes since several species were absent. However, ongoing work by Mossman and her co-workers has indicated that raising elevation within small plots and planting seedlings meets with a high degree of success. If such concepts can be scaled up to larger areas then this could help ensure that saltmarsh communities within schemes are a closer match to natural marshes.

Another issue with creating habitats within schemes in areas of low salinity is the colonisation of site by the common reed (*Pragmites australis*). This can be a problem in the upper parts of estuaries with freshwater inflows. Such issues have recently been encountered in feasibility studies looking at managed realignment options in the River Clyst, a tributary to the Exe Estuary, and the inner parts of Poole Harbour (Fig. 9.4).

9.2.2   Is There Sufficient Land to Offset the Predicted Losses of Coastal Habitats?

Coastal habitats have been lost around England and Wales due to development, conversion to agricultural land and coastal erosion. It has been estimated that extent of coastal habitats in England (excluding mudflats), declined by 13,000 ha from 1945 to 2010. This was comprised of the loss of 4,800 ha of saltmarsh, 5,000 ha of shingle and 2,600 ha of dunes (Adaptation Sub-Committee 2013). Ongoing sea level rise means that habitats will continue to be lost due to erosion and coastal squeeze. Predicting the extent of these losses is, however, subject to high levels of uncertainty. For example the Adaptation Sub-Committee (2013) stated that there were 21,000 ha of saltmarsh and 45,000 ha of mudflat at risk of coastal squeeze in front of defences in England. However, these numbers are substantially greater than those given by Defra/EA (2006), which assessed that up to 4,420 ha of saltmarsh could be lost due to coastal squeeze in England over the next 100 years. The latter report also identified a potential gain in other coastal marshes of 4,514 ha, giving a net increase of 94 ha overall. These differences illustrate the difficulties in making accurate long-term predictions of net changes in habitat extent at a national scale and in using these predictions to set appropriate targets for the amounts of compensatory habitat required.

The Adaptation Sub-Committee (2013) analysed the SMP policies for England and calculated that managed realignment policies could create up to 6,200 ha by 2030. This indicates that significant areas of our coastal zone are potentially suitable for managed realignment. This concurs with the findings of the RSPB (2002) who concluded that 33,088 ha of land were suitable for managed realignment, which was more than sufficient to ensure no net loss of inter-tidal habitats for 60 years. However, in order to create 6,200 ha by 2030 the current rate of implementation would have to double from its current average rate of 130 ha/year. Given the difficulties in assessing the future losses of habitat, it is not clear whether these predicted gains will be sufficient to compensate for all of the past and future losses, although they will certainly be a significant step in the right direction. Whether or not such large areas of realignment can actually be delivered depends on a number of factors which are discussed in the following sections.

9.2.3   Stakeholder Support

From many perspectives already discussed here, managed realignment policies can be argued to be more sustainable than traditional hard engineering solutions. However, there are a number of other aspects that make local stakeholders reluctant to welcome such schemes. Local stakeholders may lose land and property, or may need to alter their use of the land following realignment. Strong stakeholder objection may prevent schemes obtaining planning permission and thus prevent them from being implemented. When the first managed realignment schemes were implemented in the UK in the 1990s, there was little local stakeholder interest. However, public interest has grown over the last two decades and stakeholder consultation has become increasingly important. There are many reasons for this increased public interest, including: the more widespread occurrence of managed realignment, the larger size of some schemes, increased media coverage and internet access to information. The greater involvement of local stakeholders has been further encouraged by the Localism Bill[2]. This, plus the increasing involvement of third-party funders under the latest Flood and Coastal Erosion Risk Management funding mechanism, may lead to demands for schemes that meet local needs but are unsustainable in the long-term or have adverse effects further afield. This may make it more difficult to implement local managed realignments schemes (see Pontee et al. 2011).

A key aspect in the successful promotion of managed realignment schemes is therefore the early and continued engagement of local stakeholders in the decision making process. Managed realignment schemes can provide a number of benefits (e.g. bird watching, fishing, walking, cycling, horse riding and other leisure activities) which may help attract the support of local stakeholders. However, careful dialogue is needed with local residents to ensure that such interests do not lead to unwanted visitors to areas which have hitherto been quiet backwaters. Whilst managed realignments do lead to changes in agricultural land use, some practices such as low intensity grazing may continue, and discussions with landowners along these lines can help gain acceptance for schemes.

9.2.4   Costs of Managed Realignment

In England at present, the Environment Agency typically works to a guideline cost of £ 50,000 per ha for managed realignment schemes. However, many have questioned the validity of this particular value, especially since it is not index linked. A review of the cost of managed realignment over the last 20 years in the UK (Rowlands 2011) has shown that costs per hectare have varied widely depending on factors such as the size of the scheme, the promoter and the date of implementation. A recent review of the likely future costs of managed realignment in Wales shows costs varying from £ 100,000 to 675,000 per ha (Park 2013), with the higher costs for schemes on contaminated land (see also Latham et al. 2013). High land prices in some parts of the country (e.g. around the Thames estuary) can also make implementing managed realignment difficult for government agencies such as the Environment Agency, who may not be able to justify the expenditure required to purchase the land. These concerns may be less of an issue for private developers, such as ports, who may be able to better justify high costs against anticipated returns on investment.

---

[2] The Localism Bill (2011) gives more power and freedom to local authorities, communities and individuals to decide about planning local development. The legislation documentation can be accessed at: http://www.legislation.gov.uk/ukpga/2011/20/contents/enacted/data.htm.

In the right setting, managed realignment can be a cost effective coastal management measure in the long-term and this is a key reason why managed realignment policies are chosen at SMP level. The Adaptation Sub-Committee (2013) concluded that the costs of implementing managed realignment in England to 2030 (£ 10–15 million per year), would be more than offset by the financial savings on flood defence (£ 180–380 million, excluding flood storage benefits), as well as the environmental benefits (£ 80–280 million). Cost savings can arise by reducing maintenance requirements for flood defences in a number of ways. For example, managed realignment schemes may remove the requirement for flood defences all together, shorten the length of defences and/or create new defences in areas of reduced wave and tidal energy. Cost savings can also arise if schemes act as flood storage areas, thereby leading to lower water levels during storms[3] and reduced maintenance costs elsewhere (e.g. Wheeler et al. 2008).

Managed realignment may not, however, be the cheapest option in all situations. The initial costs are likely to be higher than maintaining existing defences in the short term and, compared to no active intervention, managed realignment is a costly option. On large floodplains where high land lies a substantial distance inland, creating managed realignment schemes may require construction of long defences across the floodplain, which increases scheme costs. A recent review by the Adaptation Sub-Committee (2013) reported that the overall savings in maintenance costs may have been rather overstated to date due to the failure to account for additional maintenance costs associated with new pumping stations, breach maintenance and the larger dimensions of the new defences. These factors can considerably affect the cost-effectiveness of managed realignment schemes particularly for small schemes.

9.2.5   Agricultural Land and Food Issues

In several managed realignment sites, such as Hesketh Outmarsh West (Tovey et al. 2009) and Steart (Burgess et al. 2013), there is a drive to continue low intensity grazing. This does not however, deal with the issue of losing areas of high grade arable farmland, food security and land prices in relation to managed realignment (Walsh 2008). Additionally, increasing grain prices have driven up the market value of agricultural land in the UK over the last decade and further increases could make managed realignment more difficult in some areas in the future.

The Adaptation Sub Committee (2013) concluded that managed realignment would only result in the loss of 0.1 % of England's high grade agricultural land by 2030. They also concluded that these losses could be offset by taking advantage of the other ecosystem services and opportunities provided by managed realignment, such as aquaculture, fish nursery and grazing. This conclusion appears to suggest that the loss of agricultural land will not pose a problem in the implementation of

---

[3] Note: In some instances managed realignment can also lead to water levels being raised within estuaries (see Burgess et al. 2013).

more realignment in the future. However, not all farmers will be willing to embrace such changes and substantial stakeholder engagement is likely to be required both nationally, with the National Union of Farmers; and locally, with individual farmers. An incentive for farmers to sell their land for managed realignment schemes may be the payment of an increment on the market value of the land based on the additional ecosystems services that may be gained.

9.2.6 *Mitigation for Freshwater Habitat Displacement*

Managed realignment schemes that lead to a loss of internationally designated freshwater habitats (e.g. Special Protection Areas, Special Areas of Conservation, Ramsar sites) have to allow for the re-creation of these habitats elsewhere. This raises the question of whether there is sufficient space to create these freshwater habitats further landwards. This can be particularly challenging in estuaries where the current balance of freshwater and marine habitats may have been significantly altered by embankment construction over many centuries (see for example Pontee et al. 2013). However, to date freshwater compensation issues have been dealt with on a site-by-site basis and, whilst adding difficulty to some schemes, have not prevented the uptake of managed realignment at a national level. Nevertheless, studies to determine the availability of secondary compensation freshwater sites at a national scale would seem to be necessary to inform any assessment of the long-term sustainability of managed realignment as a policy.

9.3 Conclusions

Managed realignment can offer a sustainable long-term management option for coasts and estuaries by reducing the pressures on flood defences and compensating for habitat lost due to developments or coastal squeeze. The habitats created also provide other ecosystem services and opportunities, such as: aquaculture, fish nursery, grazing and recreation.

Where schemes are undertaken to compensate for the loss of specific habitat types then careful scheme design is needed. It can be difficult to create sustainable mudflat in estuaries with high suspended sediment concentrations and although saltmarsh readily forms in managed realignment sites, some rare species may be absent. Additionally *Phragmites* can prove an unwanted colonising plant in some low salinity settings. These difficulties, plus the fact that many estuaries currently have an artificial distribution of designated habitats, suggest that future habitat management could usefully consider wider regions or biogeographic areas in order to allow some 'natural' changes in the balance of habitats in some estuaries or European Sites (see Pontee et al. 2013).

Over 10 years ago, Defra (2003) noted that managed realignment was perceived as a long- rather than a short-term solution, due to it being politically less acceptable then. Undoubtedly there will always be some locations where managed realignment cannot be promoted in the short-term. However, the implementation of more than 50 schemes in the last two decades indicates that, at a national scale, managed realignment can be delivered in relatively short-time.

To create all the managed realignment schemes identified in SMPs by 2030 will require a doubling in the current rate of implementation. Studies have shown that there is more than enough suitable land on which to create managed realignment schemes around our coasts (RSPB 2002; Adaptation Sub Committee 2013). However, managed realignment is not the best option for all coastal areas. Policies of managed realignment for flood risk management alone may be uneconomic in areas of high land values, where long lengths of new defences are needed or where there is contaminated land. Increases in agricultural land prices, plus competition between private developers and Government Agencies, may also make managed realignment schemes more difficult to implement in the future. Additionally, local stakeholder objections may prevent schemes from gaining planning permission in some areas.

In the future, it is certain that we will continue to see managed realignment schemes being built in the UK. However, managed realignment is not a universal panacea—it will not be possible to implement this policy everywhere, and schemes are unlikely to be viewed as a success from all perspectives. Increases in third party funding and greater local involvement in decisions mean that managed realignment schemes will require the buy-in of many stakeholders. The continued implementation of managed realignment schemes in the future will therefore require extensive stakeholder consultation.

References

ABP Research and Consultancy Ltd. (1998). Review of coastal habitat creation, restoration and recharge schemes (Report No. R 909), p. 78. Southampton: ABP Research & Consultancy Ltd.
Adaptation Sub-Committee. (2013). Progress report 2013—managing the land in a changing climate. Committee on Climate Change, p. 136. http://www.theccc.org.uk/publication/managing-the-land-in-a-changing-climate/. Accessed 29 Dec 2013.
Atkinson, P. W., Crooks, S., Grant, A., & Rehfisch, M. M. (2001). The success of creation and restoration schemes in producing intertidal habitat suitable for waterbirds. English Nature Research (Report No. 425). English Nature, Peterborough.
Burgess, K., Pontee, N., Wilson, T., Lee, S. C., & Cox, R. (2013). Steart Coastal management project: Engineering challenges in a hyper-tidal environment. Coasts, Marine Structures and Breakwaters 2013, Aberdeen, 17–20 September 2013.
Canning, P., & Pontee, N. (2013). Assessing habitat compensation requirements in estuary environments. Coasts, Marine structures and breakwaters 2013, Aberdeen, 17–20 September 2013. Institution of Civil Engineering. http://www.ice.org.uk/ICE_Web_Portal/media/Events/Breakwaters%202013/Assessing-Habitat-Compensation-Requirements-in-Estuary-Environments.pdf. Accessed 8 June 2014.

Defra. (2003). Managed realignment review. Defra/Environment Agency Policy Theme R&D Project FD2008, Defra Publications, Defra-Flood Management Division, Ergon House, London SW1P, 2003, 2AL. p. 222 + App.

Defra. (2006). Shoreline management plan guidance. Volume 1: Aims and requirements, March 2006. https://www.gov.uk/government/publications/shoreline-management-plans-guidance. Accessed 29 Dec 2013.

Defra/EA. (2006). National evaluation of the costs of meeting coastal environmental requirements NEoCOMER). R&D technical report FD2017/TR Joint Defra/EA flood and Coastal erosion risk management R&D programme. Report produced by Risk and Policy Analysts Ltd, Royal Haskoning UK Ltd and ABP Marine Environmental Research Ltd for Defra/EA, April 2006, p. 2.

European Commission. (1992). 'Habitats Directive'—Council directive 92/43/EEC on the conservation of natural habitats and of wild fauna and flora. European community directive adopted in 1992.

European Commission. (2009). Council Directive 2009/147/EC on the conservation of wild birds (the codified and consolidated version of council directive 79/409/EEC as amended).

Latham, D., Pontee, N., Monroe, B., & Bond, I. (2013). Reconstructing habitats in a heavily industrialised estuary: From brine wells to salt marsh. Coasts, Marine structures and breakwaters 2013, Aberdeen, 17–20 September 2013.

Mazik, K., Musk, W., Dawes, O., Solyanko, K., Brown, Su., Mander, L., & Elliott, M., (2010). Managed realignment as compensation for the loss of intertidal mudflat: A short term solution to a long term problem? *Estuarine Coastal and Shelf Science, 90*(1), 11–20.

Morris, R. K. A. (2012). Managed realignment: A sediment management perspective. *Ocean & Coastal Management, 65,* 59–66.

Mossman, H. L., Davy, A. J., & Grant, A. (2012). Does managed coastal realignment create saltmarshes with 'equivalent biological characteristics' to natural reference sites? *Journal of Applied Ecology, 49*(6), 1446–1456.

Nottage, A., & Robinson, P. (2005). The saltmarsh creation handbook: A project manager's guide to the creation of saltmarsh and intertidal mudflat. London: RSPB, Sandy and CIWEM.

Park, R. (2013). *Challenges in Wales*. Presentation given at the ABPmer Habitat Creation Conference, 20th November, 2013. London: Institution of Civil Engineers. http://www.abpmer.net/omreg/. Accessed 29 Dec 2013.

Pontee, N. (2014). Accounting for siltation in the design of intertidal creation schemes. *Ocean and Coastal management, 88,* 8–12.

Pontee, N. I. (2003). Designing sustainable estuarine intertidal habitats. Engineering sustainability. *Proceedings of the Institution of Civil Engineers, 156*(Issue ES3), 157–167.

Pontee N. I., & Parsons, A. (2010). A review of coastal risk management in the UK. Proceedings of the institution of civil engineers. *Maritime Engineering Journal, 163*(MA1), 31–42.

Pontee, N. I., & Parsons, A. (2012). Adaptation as part of sustainable shoreline management in England and Wales. Proceedings of the institution of civil engineers. *Maritime Engineering Journal, 165*(MA3), 113–130.

Pontee, N. I., Hamer, B. A., & Turney, P. D. (2005). *Sustainable flood risk management in European estuaries*. Proceedings of the 1st international conference on coastal management and engineering in Middle East (ARABIAN COAST 2005), Dubai, 27th–29th November 2005. Theme D: COASTAL RISK MANAGEMENT. http://www.iahr.net/site/e_library/ConfPro/ArabianCoast2005/ArabianCoast2005_ToC.pdf. Accessed 4 Jan 2014.

Pontee, N. I., Parsons, A., & Ashby Crane, R. (2011). Local empowerment: A barrier to strategic flood and coastal risk management. Abstract for presentation at: Water & Environment 2011: CIWEM's annual conference, big society—future environment, 06 Apr 11–07 Apr 11, Olympia, London.

Pontee, N., Ashby Crane, R., & Batty, L. (2013). A fresh look at managed realignment: Estuary wide and long term sustainability. Coasts, Marine structures and breakwaters 2013, Aberdeen, 17–20 September 2013. http://www.ice.org.uk/ICE_Web_Portal/media/Events/Breakwaters%202013/A-Fresh-Look-At-Managed-Realignment-Estuary-Wide-and-Long-Term-Sustainability.pdf. Accessed 1 Feb 2014.

Roberts, T. H. (1991). Habitat value of man-made coastal marshes in Florida. US Army Corps of Engineers, Waterways Experiment Station, 1991, Wetlands Research Programme Technical Report WRP-RE-2, Mississippi.

Rowlands, O. E. (2011). Current trends in the costs of managed realignment and regulated tidal exchange. Schemes in the UK. Unpublished MSc thesis, University of Southampton, Southampton, UK, p. 81.

RSPB. (2002). Seas of change—the potential area for inter-tidal habitat creation around the coast of mainland Britain. Authors: Pilcher, R., Burston, P., Kindleysides, D., & Davis., R. October 2002, p. 35.

Toft, A. R., & Madrell, R. J. A. (1995). Guide to the understanding and management of saltmarshes. National Rivers Authority, Project 444, pp. 1–213.

Tovey, E. L., Pontee, N. I., & Harvey, R. (2009). Award winning managed realignment: Hesketh out Marsh West. *Target Journal: Proceedings of the Institution of Civil Engineers, Sustainable Engineering, 162*(4), 223–228.

Walsh, K. (2008). Food fight. New civil engineer, 2008. Issue dated 109th April 2008 'Food fight', pp. 14–15. http://www.nce.co.uk/food-fight/1098084.article. Accessed 13 Jan 2014.

Wheeler, D., Tan, S., Pontee, N., & Pygott, J. (2008). Alkborough scheme reduces extreme water levels in the Humber Estuary and creates new habitat. FLOODrisk 2008—the European conference on flood risk management research in to practice 30 September–2 October 2008 Keble College, Oxford, UK.

Chapter 10
Current Perceptions About Managed Realignment

Luciana S. Esteves

Abstract Despite the diversity of managed realignment approaches adopted in different countries, there are certain elements of commonality that provide relevant lessons that can be applicable to facilitate and improve the implementation of managed realignment in general. Public perception and stakeholders' engagement are often cited as a factor limiting the wider implementation of managed realignment. Considering the scale and importance of managed realignment in national and regional strategies in many countries, gaining wider public acceptance is fundamental. To take adequate actions that can improve social uptake of managed realignment, it is first necessary to understand the differences between public perception and the current knowledge of practitioners and researchers. This chapter describes the current thinking of researchers and practitioners about the potential, performance and limitations of managed realignment and contrasts these with the perception of stakeholders. Critical differences are identified and ways to address these with the aim to increase the social acceptance of managed realignment are suggested.

The previous chapters provide an overview of the range of approaches and the diversity of context in which managed realignment has been implemented in Europe and elsewhere. Differences in physical settings, objectives, technical approaches, and political, institutional and cultural background make existing managed realignment projects almost unique. Many differences also exist across national and regional strategies (see Chap. 4). It is therefore difficult to generalise achievements and limitations as these vary from case to case and depend on the intended objectives and general expectations. Despite these many variations, there are certain elements of commonality that provide relevant lessons that can be applicable to facilitate and improve the implementation of managed realignment in general.

Public perception and stakeholders engagement are often cited in this book (e.g. Chaps. 2, 5, 8 and 9) and in the wider literature (e.g. McGlashan 2003; Myatt et al. 2003a, b; Ledoux et al. 2005; French 2006; Goeldner-Gianella 2007; Roca and Villares 2012) as a factor limiting the wider implementation of managed realignment. Considering the scale and importance of managed realignment in national

L. S. Esteves (✉)
Faculty of Science and Technology, Bournemouth University, Talbot Campus,
Poole, Dorset, BH12 5BB, UK
e-mail: lesteves@bournemouth.ac.uk

L. S. Esteves, *Managed Realignment: A Viable Long-Term Coastal Management Strategy?*, 109
SpringerBriefs in Environmental Science, DOI 10.1007/978-94-017-9029-1_10,
© Springer Science+Business Media Dordrecht 2014

and regional strategies of climate change adaptation, habitat creation and flood risk management, gaining wider public acceptance is fundamental. To take adequate actions that can improve social uptake of managed realignment, it is first necessary to understand the differences between public perception and the current knowledge of practitioners and researchers. It is timely to identify the key factors leading to the negative social perception about managed realignment.

Research on social perceptions about managed realignment is limited (Roca and Villares 2012) and dominantly concerns projects in the UK (e.g. Myatt-Bell et al. 2002; Myatt et al. 2003a, b; Midgley and McGlashan 2004; Parrott and Burningham 2008), with some recent studies related to future projects elsewhere (e.g. SOGREAH 2011; Roca and Villares 2012). Assessment of social perception includes reactions to proposed projects (Myatt et al. 2003a; Midgley and McGlashan 2004; SOGRE-AH 2011; Roca and Villares 2012), response to implemented projects (e.g. Myatt-Bell et al. 2002; Myatt et al. 2003b) and general views about implementation and performance of managed realignment (e.g. Ledoux et al. 2005; Goeldner-Gianella 2007; Parrott and Burningham 2008). However, none of the existing studies have assessed how the wider social perception contrasts with the understanding of researchers and practitioners directly or indirectly involved in managed realignment.

An online survey was conducted in 2013 with the objective of obtaining a general overview of current perceptions about managed realignment. To obtain a wide range of participants from the UK, Europe and elsewhere, a link for the survey was distributed via twitter (to relevant groups and organisations), professional online discussion groups (e.g. in LinkedIn) and email sent to researchers and practitioners. A total of 238 usable responses were obtained, 63% from the UK and 47% from a wide range of geographical spread (see Table 10.1). The geographical spread reflects well the distribution of experiences in managed realignment to date (when considering the most common use of the terminology) and therefore results can be considered a valid representation of current overall perceptions.

Additionally, the range of UK participants allowed contrasting the views of practitioners, researchers and stakeholders. Results concerning the views in the UK are partly presented in Esteves and Thomas (2014). This chapter summarises the main findings of this study and extends the analysis to include all responses. The results of the online survey and findings from existing relevant literature are then used to identify the key issues that need to be addressed to improve social acceptance and uptake of managed realignment (Table 10.2). It is anticipated that these key findings are applicable to inform future policy developments concerning managed realignment in the UK and elsewhere.

10.1 Suitability of Managed Realignment for Habitat Creation and Flood Risk Management

The great value of managed realignment as a more sustainable coastal management option is related to its multiple functions and the ecosystem services the created habitats can provide (e.g. Luisetti et al. 2011; Spencer and Harvey 2012). As

Table 10.1 Characteristics of the respondents in the online survey '*Your views about managed realignment*'

Total *n*=238	UK (*n*=151)	Other (*n*=87)
Represented sectors		
Private consultants Stakehold-	29%	23%
ers Researchers/academics	22%	3%
Government	21%	56%
NGO	14%	15%
	14%	3%
Geographical distribution	35% East England	37% Europe
	26% South England	16% USA and Canada
	17% Southwest England	15% Oceania
	8% Northeast England	5% Africa
	6% Scotland	6% Latin America
	4% Wales	8% Other
	4% Northwest England	13% Unknown
Type of involvement		
Not involved	15%	25%
Flood risk management Habitat	71%	53%
compensation Creation of	70%	40%
habitat	78%	38%

a result, most projects have multiple objectives including: habitat creation; compensation for habitat loss; improvement of flood defences; and reduction of costs to maintain defences (e.g. Esteves 2013). Results from the online survey indicate that the majority of respondents (56%) agree that managed realignment is a good mechanism to deliver sustainable flood risk management with added environmental benefits. About 12% of respondents disagree with this statement and these are mainly stakeholders and members of the public. Concerning the same statement, Esteves and Thomas (2014) show that around 60% of practitioners, consultants and researchers in the UK expressed agreement, contrasting with only 9% of stakeholders.

Similar results are found about whether managed realignment is a promising strategy to reduce flood risk and the costs to maintain coastal defences (Fig. 10.1b). However, overall agreement is expressed by about 45% of respondents, with more positive answers from non-UK than UK respondents (57% and 38% respectively). A contrasting 76% of stakeholders and members of the public in the UK disagree with the statement shown in Fig. 10.1b, comparing with only 14% of practitioners and consultants (Fig. 10.1b, bottom).

Opinions are more divided about the statement that managed realignment is better suited for the creation of intertidal habitats than for flood risk management (Fig. 10.1a). In the UK, stakeholders and researchers responded similarly, roughly showing equally divided opinions, while 61% of consultants and 41% of practitioners disagreeing with the statement (Fig. 10.1a, bottom). Results indicate that the majority of respondents are satisfied with the results from managed realignment projects. About 54% of all respondents disagree and 14% agree with the statement

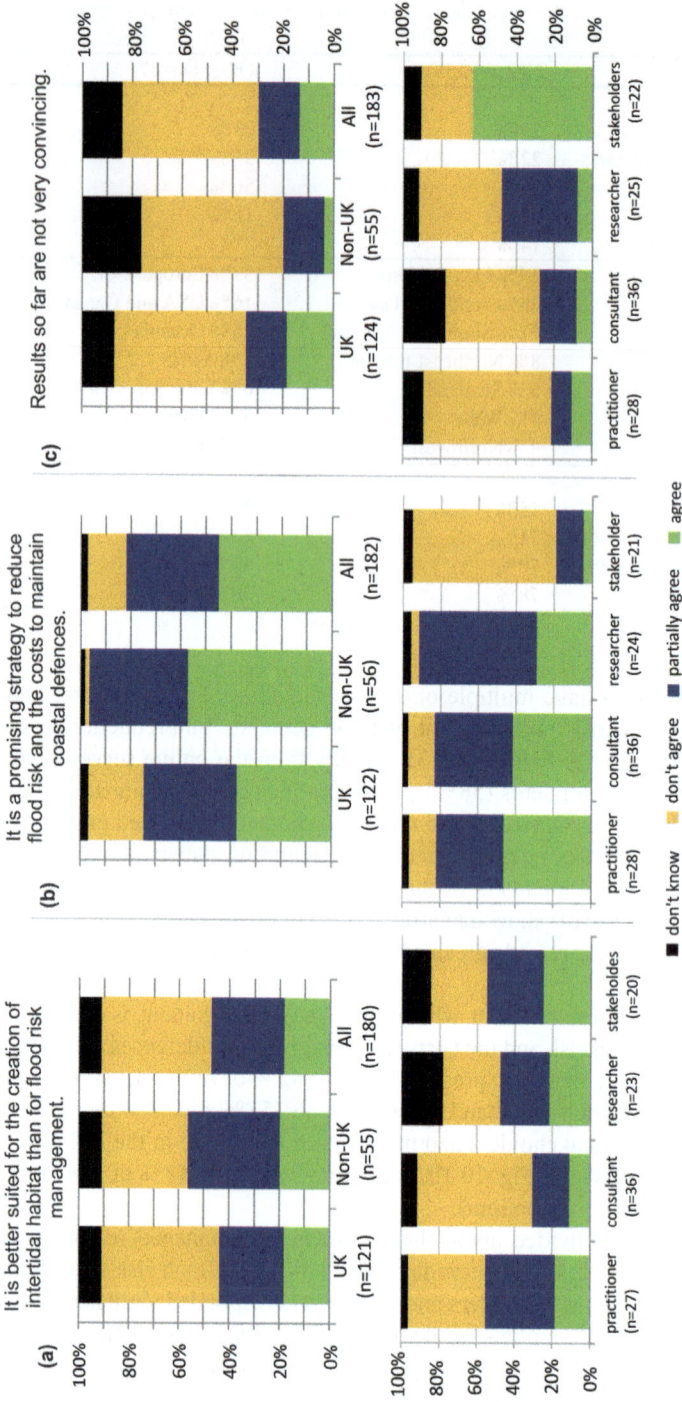

Fig. 10.1 Results (in percentage of responses) obtained from the online survey on current views about specific statements related to managed realignment, as shown in (**a**), (**b**) and (**c**). Responses from UK, non-UK and all participants are show at the *top*. The views of practitioners (government and non-government), consultants, researchers and stakeholders (including members of the public) from the UK are shown at the *bottom* (modified from Esteves and Thomas 2014)

that results are not very convincing so far (Fig. 10.1c). The perception of stakeholders in the UK once again contrasts with other respondents (Fig. 10.1c, bottom) and contribute to increase the percentage of negative responses.

In general terms, respondents support the concept that managed realignment is a good strategy for managing flood risk and deliver wider environmental benefits. However, it is concerning that stakeholders' responses are strikingly more negative when compared with the other groups. It was expected that consultants and practitioners would have a more positive view about managed realignment as they are involved in project design and delivery. Nevertheless, there is a considerable proportion of partial agreement with the statements offered in the survey, which may indicate variability of results across sites and uncertainty about the progress of sites recently implemented (see Sect. 10.2).

Negative comments from stakeholders often express that loss of land is not being balanced by a clear benefit, and there is a lack of trust in the parties involved in policy, planning and implementation of managed realignment (a limitation also identified by Roca and Villares 2012). A clear and consistent participatory process is necessary to establish confidence amongst the parties involved (e.g. Midgley and McGlashan 2004; Ledoux et al. 2005) that may lead to the development of working partnerships essential for the viability of managed realignment (see Chap. 8).

10.2   Performance of Projects

Figure 10.2 shows that opinions about the performance of managed realignment projects against the planned objectives are similar in the UK and elsewhere. It is clear that the majority of respondents think projects are performing well in all or some aspects and only a small proportion of respondents think projects are not performing well. Outside the UK, views tend to be slightly more positive, with lower proportion of respondents indicating that projects are not performing well so far and a higher proportion thinking projects are performing well.

The differences between UK and non-UK responses are due mainly to the views of stakeholders and members of the public. Only a few of these participants are from outside the UK and they have not answered the question shown in Fig. 10.2. On the other hand, the members of the public and stakeholders from the UK expressed a more negative view than practitioners and researchers (Fig. 10.2b). Considerably larger percentage of stakeholders (including members of the public) think projects are not performing well in comparison with other respondents (note that none of the practitioners indicated projects are not performing well so far). Not surprisingly practitioners (who are directly involved in implementation and management of projects) show the most positive views with 36 % of respondents indicating projects are performing well in all objectives (Fig. 10.2b).

A good proportion of respondents from all groups think projects are performing well in some aspects and not so well in others. It is understandable that sites are perceived to be performing differently due to a number of reasons. Respondents

Based on your understanding, how are the managed realignment projects you have
been involved with performing against the intended objectives?

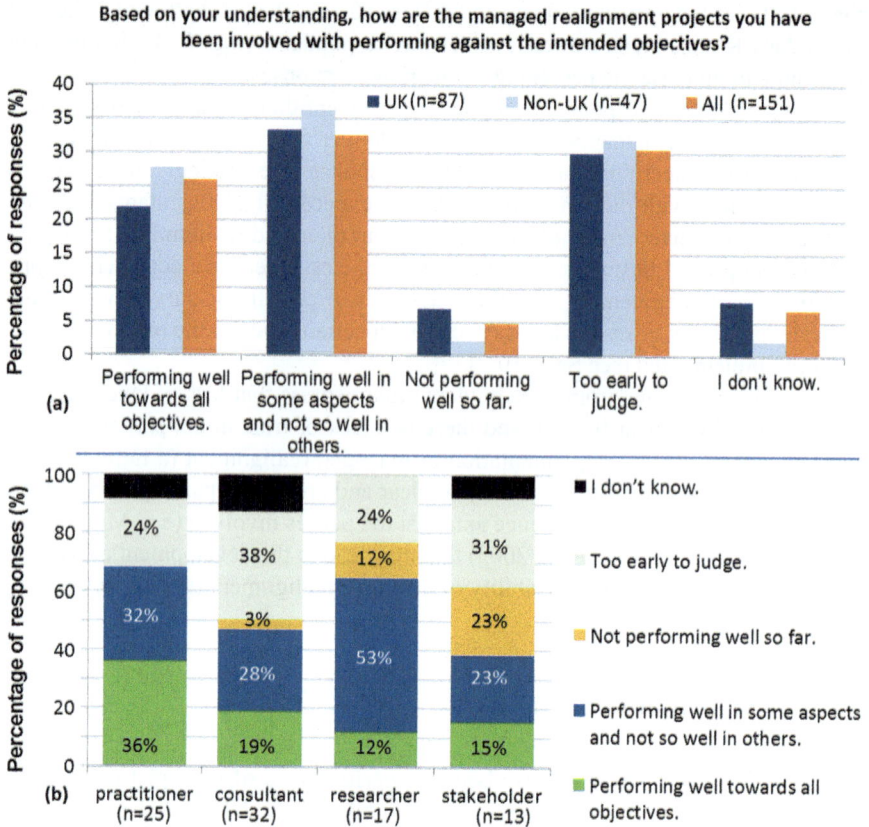

Fig. 10.2 Results obtained from the online survey on current views about the performance of managed realignment projects. (**a**) Comparison of responses from UK, non-UK and all participants. (**b**) The views of practitioners (government and non-government), consultants, researchers and stakeholders (including members of the public) from the UK (modified from Esteves and Thomas 2014)

indicate that many projects are at early stages of implementation and therefore too early to judge performance of all or some aspects. Others indicate that performance varies from poor to good across projects, while performance cannot be measures in some cases due to the lack of set objectives. Poor or variable results have also been attributed to issues related to project design (e.g. breach to small; poor drainage; land elevation too low or too high) or insufficient funds to properly implement the planned design and/or monitor progress.

About 30 % believe it is too early to judge and this is expected as many realignment projects have been implemented in the last 10 years and they are at different stages of development. Perceptions may change through time depending how sites evolve and whether projects have clear set targets (so performance can be assessed). Only time will tell to which direction the outcome of these projects will shift perceptions.

If you are not located in the UK, can you please indicate the level or awareness and
experience about managed realignment in your country (to the best of your knowledge):

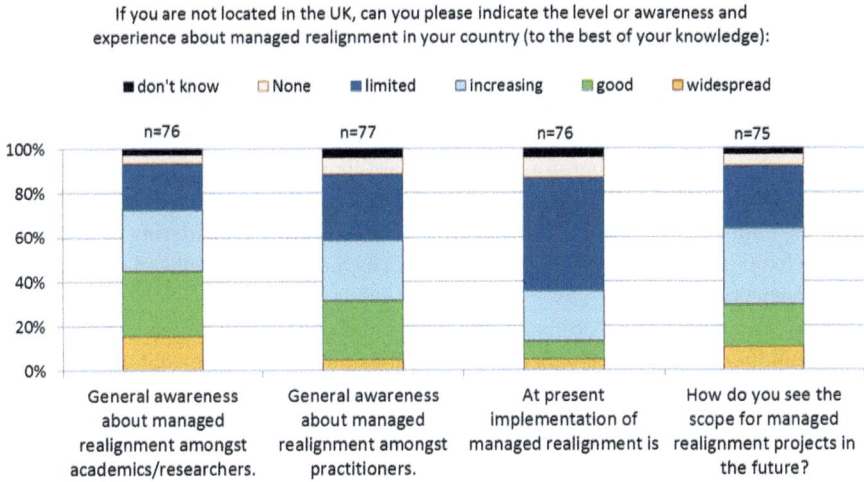

Fig. 10.3 Online survey results concerning current awareness, implementation and future potential for managed realignment outside the UK

10.3   Current Perceptions About the Future Scope for Managed Realignment Outside the UK

Figure 10.1 shows the result of a question aimed to obtain an overview of current perceptions about awareness, implementation and future potential for managed realignment outside the UK. It is noticeable (although expected) that implementation of managed realignment is considered limited at present (by 52 % of respondents); however, the answers indicate that managed realignment might be more widespread in the future. About 35 % of respondents say that implementation of projects at present are widespread, good or increasing, while 64 % say the same concerning implementation of managed realignment in the future. Not surprisingly, results indicate that currently researchers are more aware about managed realignment than practitioners but 27 % of respondents say awareness is increasing in both sectors (Fig. 10.3).

Current awareness and implementation of managed realignment is said to be widespread or good and will continue growing in the future by respondents from Belgium and Canada. Although information of geographical location was not always provided, it is possible to assess perceptions of trends in some countries. Respondents from Australia, New Zealand and India indicate that implementation of managed realignment is limited at present in their countries but likely to increase in the future. Based on the existing literature (Alexander et al. 2012; Niven and Bardsley 2013; Taylor and McAllister 2013; Harman et al. 2014; Reisinger et al. 2014), it is likely that Australia and New Zealand are focusing more on managed retreat than other forms of managed realignment. In the USA and France there are divided opinions on whether implementation is currently limited or increasing, but it is clear that most anticipate a step up in managed realignment experiences in the future in comparison to the present. On the contrary, respondents from Portugal, Spain and the Netherlands indicate that currently experiences in their country are limited and will continue to be limited in the future.

It is important to note that a definition of managed realignment was not provided in the survey and therefore responses are likely to reflect the most common interpretation adopted in their region. Some open comments indicate, for example, that respondents from the Netherlands and France are not considering methods of controlled tidal restoration (i.e. CRT and RTE) as methods of managed realignment when answering the questions.

In France, Spain and Portugal restoration of tidal flow into embanked areas occurs mainly as a result from accidental breaches due to storms or natural decay of abandoned defences (e.g. Cearreta et al. 2013; Almeida et al. 2014). In many cases, there is the view that restoration of natural habitat under these conditions produces similar results to managed realignment (e.g. Cearreta et al. 2013). The literature suggests that the vegetation structure of restored saltmarshes is likely to differ from natural systems (Zedler and Adam 2002; Wolters et al. 2005; Spencer et al. 2008; Mossman et al. 2012), even after time-frames of 50–100 years (Garbutt and Wolters 2008; Mossman et al. 2012). This difference raises the question whether restored saltmarshes are able then to deliver equivalent ecosystems services and functions (e.g. Craft et al. 2003; Garbutt and Wolters 2008; Santín et al. 2009; Mossman et al. 2012; Burden et al. 2013). Until knowledge is advanced to clarify this issue, it is up to debate whether there is justification for the expenditure of public funds in managed realignment projects. Evidence that tax payer money is being used to provide tangible wider benefits (e.g. improvement of flood risk management, recreation or opportunities for economic activities) is required to shift perceptions.

Comments from respondents indicate that the wider implementation of managed realignment is limited by insufficient number of demonstration projects and opposition from stakeholders and the public. The UK experience is often taken as guidance but differences in physical settings, government policies/structure and cultural background are not always applicable. Lack of understanding the long-term evolution of sites is also cited as a limiting factor (see Sect. 10.4). This stems from the fact that only few examples exist that are more than a decade old and not much is known about their evolution and the ecosystem services they are providing.

It is clear from the results shown in Fig. 10.1 that there is room for improving awareness of researchers and especially practitioners about managed realignment. Practitioners need to be knowledgeable about existing experiences and convinced that managed realignment is the best option to promote the idea at the local level. The amount of time and effort involved in public engagement required to increase awareness and change perceptions about managed realignment is another issue raised by respondents.

The wider implementation of managed realignment outside the UK is faced with a conundrum. More evidence and examples are required to increase the understanding of practitioners and researchers about managed realignment and facilitate public acceptance. However, opportunities for implementation of managed realignment projects are limited because of the novelty of the concept. It is therefore key that robust evidence obtained from monitoring of existing projects (within and outside the context of the country) is widely disseminated to inform practitioners and the public.

The limiting factors identified above are not too dissimilar of the ones encountered in the UK. However in the UK, many communities and stakeholders have already had direct or indirect involvement with existing projects. Not only are more people aware of the terminology associated with managed realignment, social perceptions are already influenced by the way projects were implemented and public engagement conducted. There is room for improvement in many aspects of the planning and implementation process; more importantly it is time that uncontested and clear evidence of realised benefits is widely disseminated. To increase reliability and transparency it is important that evidence is assessed independently and based on robust data analyses, which should address individual projects and overall achievements of regional and national targets.

Evidence is required to: improve project design, inform policy development, educate practitioners and attract support from stakeholders and the public. Practitioners are unlikely to convince stakeholders and the public if they are uncertain about the benefits (and losses) brought by managed realignment.

10.4   Main Limitations

Managed realignment is a novel and challenging concept to most communities and great efforts are required to gain support for project implementation at the local level. It has been suggested that wider acceptance might increase the longer the concept is being discussed and more projects implemented (Pethick 2002; Myatt-Bell et al. 2002; Myatt et al. 2003a; b). Only when managed realignment becomes a more established coastal management approach, there might be enough public understanding and acceptance to influence opinions at the local level. However the 'not in my backyard' attitude might still prevail irrespective of general increased acceptance.

It is clear from the results of the online survey presented so far that there is a considerable difference in the perception of stakeholders about the suitability and performance of managed realignment projects in comparison with other groups involved in managed realignment. Similar discrepancy is found about the factors limiting the implementation of managed realignment. Results based on the views of all respondents participating in the online survey are show in Fig. 10.4a, while the views of stakeholders in the UK are shown in Fig. 10.4b.

In general terms, it is fair to say that overall views are divided as the majority (>50%) of respondents expressed the same view in only few cases. Additionally, only in one aspect an agreement was observed between the majority of all groups of respondents (Fig. 10.4). Over 67% of all respondents and 71% of UK stakeholders agree that better understanding about long-term evolution of the realigned sites

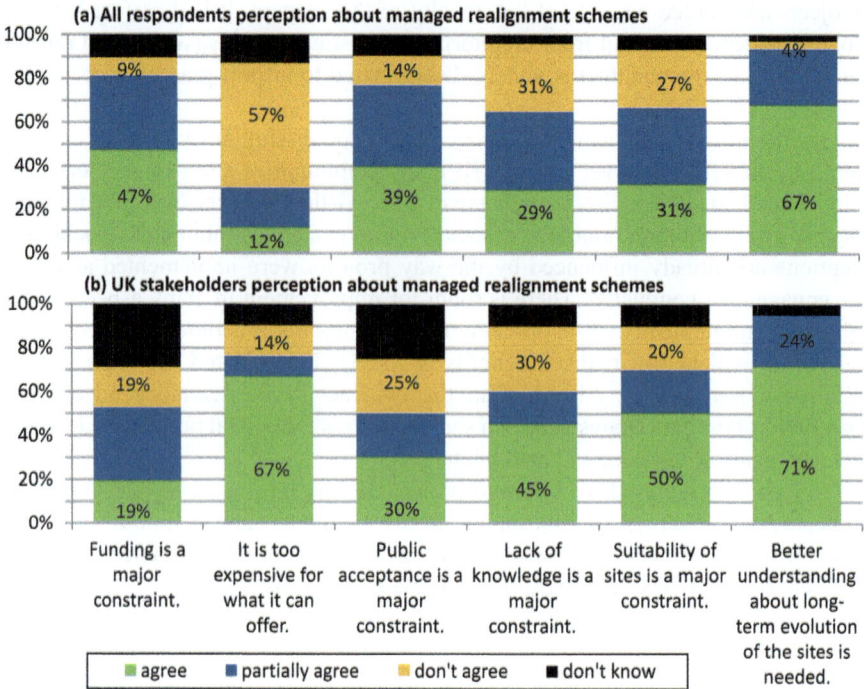

Fig. 10.4 Results of the online survey concerning factors limiting the implementation of managed realignment based on the views of (**a**) all respondents and (**b**) stakeholders in the UK

is needed (Fig. 10.4). From the practitioners and consultants perspective, there is certain confidence in the tools available to assist project design and engineering; however; it is recognised that improvement of hydrodynamic modelling capabilities and/or uptake of model results concerning sediment processes is required (e.g. Esteves and Thomas 2014). Across the groups, there is a general agreement that there is a need to better understand the changes in hydrodynamic and sedimentary processes due to managed realignment (e.g. Spencer and Harvey 2012) as they influence colonisation by the biota (e.g. Marquiegui and Aguirrezabalaga 2009; Davy et al. 2011), effects on flood risk (e.g. Morris 2012) and some biogeochemical aspects (e.g. Adams et al. 2012; Burden et al. 2013) important for the capacity to store carbon, cycle nutrients and support fisheries. Long-term monitoring of sedimentary processes is fundamental to quantify environmental changes and assess whether site evolution is occurring towards the desired objectives. Long-term, in this case, should be defined by the time the site reaches equilibrium, i.e. when trends and rates of changes become more constant.

About the same proportion of UK stakeholders and all respondents disagree that lack of knowledge is a major constraint. However, a much larger percentage of stakeholders agree with this statement (Fig. 10.4). The online survey also asked opinions about the duration and coverage of systematic monitoring at realigned sites and respondents, including stakeholders indicated they think all sites should be monitored and for longer than current general practice. Notably, the majority of researchers and stakeholders indicate that vegetation, sedimentation and biogeochemical processes should be monitored for at least 10 years or longer (Esteves and Thomas 2014). Consultants and practitioners more often think the minimum monitoring could be 5 years or less.

Opinions were generally divided concerning the statement "public acceptance is a major constrain" (Fig. 10.4). More respondents agree than disagree with the statement in both groups but interestingly more stakeholders disagree. It is possible that some stakeholders (and other respondents) have disagreed with the statement as they might believe that public opposition may delay but not stop managed realignment projects; therefore public acceptance, although important at the local level, is not considered a major constraint. Researchers are the group that most agree that public acceptance is a major constraint (42 %), but this might reflect findings from the literature rather than practical experience.

Concerning funding for and costs of managed realignment, opinions of stakeholders contrast with the perceptions of all respondents. Figure 10.4 shows that only 19 % of stakeholders think funding is a major constraint; however, 67 % perceive managed realignment to be too expensive for the benefits it may create. In contrast 47 % of all respondents agree that funding is a major constraint and only 12 % think it is too expensive. It is evident that stakeholders are more likely to believe that enough money is available for managed realignment (so funding is not an issue), but they question whether the funds are being well-spent. On the other hand, practitioners (who have to seek funding sources for the projects), are more likely to perceive funding to be an issue.

Availability of suitable areas might affect the cost of managed realignment in the future as suitable land becomes scarcer. In England, for example, the government plans to realign a total of 111 km by 2016 and 550 km by 2030 resulting in the creation of 6,200 ha of intertidal habitat at a cost of £ 10–15 million per year (Committee on Climate Change 2013). The estimated cost would not be sufficient even to maintain the current needs without the increased rate required to meet the future targets. In England (and other countries, such as Belgium), land purchase is negotiated case-by-case and subjected to high price variability. Often projects are implemented in areas behind flood defences in poor state of conservation, where land usually have a discounted value (e.g. £ 5,000/ha in Wallasea Island) and landowners are more willing to sell or work in partnership.[1] In the future, suitable land may be restricted to high grade agricultural land, which have a current average value in the UK of £ 18,400/ha (RICS 2013). Demand for land is high near ports and

---

[1] In many countries, property rights might be an issue limiting government to acquire the required land for managed realignment. In the UK, the Crown Estate takes ownership of newly intertidal areas created naturally with immediate effect regardless of who originally purchased the land. Therefore, intertidal land created by accidental breaches of defences immediately revert to Crown Estate; resulting in loss of ownership. However, if the area is created artificially (and would not be inter tidal without the intervention) then the Crown Estate will not claim ownership until such time as the area would have become intertidal naturally.

Table 10.2 Requisites for facilitating the implementation and wider uptake of managed realignment

Governance and high-level strategy	Horizontal and vertical integration of policies (e.g. integrated planning and flood risk management; clear definition of roles and responsibilities)
	Political will to take decisions for long-term and wider benefits instead of personal gain (e.g. think beyond the election cycles)
	Clear and well-justified strategic vision (targets and time-frames are widely disseminated and understood)
	Funding and institutional mechanisms that facilitate the implementation of the strategy across the national to local levels (e.g. to purchase land; to fund dissemination and educational plans)
	Ensure the mechanisms to implement the strategy do not conflict with private property rights
	Capacity building for practitioners and policy-makers related to key knowledge concepts, uncertainties and potential socio-economic implications
	Robust public dissemination (i.e. consistent message) and stakeholder engagement strategy
	Adaptive management (i.e. based on regular assessments)
Delivery at the project level	Clear targets and well-defined time-frames for each project
	Strong knowledge basis about potential benefits accruing from local projects and associated uncertainties
	Availability of suitable land to deliver regional and local targets
	Institutional capacity and expertise at the local level (or external support to enable knowledge transfer)
	Emphasis on multipurpose functions and benefits to attract wider support
	Structured strategy to overcome strong sectorial views or individual strong voices (local power)
	Robust stakeholder engagement to understand local needs and expectations
	Tailored project design to maximise benefits relevant to local communities
	Designed based on modelling outputs considering worst-case scenarios of meteorologic and oceanographic conditions
	Better understanding of long-term evolution of realigned sites and hydrodynamics and sediment interactions (e.g. evidence based on monitoring)
	Systematic monitoring of relevant parameters until rates of change/conditions stabilize
	Independent and science based data analysis to provide evidence of performance
Public perception and stakeholders engagement	Good understanding of national, regional and local targets
	Focus on multiple-functions and benefits (to attract wider interests and reduce not-in-my-backyard attitude)
	Increase trust in government and non-government players (e.g. through a transparent decision-making process and a legitimate participatory process)
	Bottom-up approach to determine local targets
	Education efforts to reduce negativity associated with 'give in to the sea' perception
	Increased awareness about ecosystem services, climate change adaptation needs, the concept of managed realignment
	Long-term dissemination and engagement plan to reduce the 'novelty effect' (i.e. to establish the concept into the public mind)
	Dissemination of evidence about the effects on flood risk to people and property and wider benefits
	Working with the media to disseminate consistent messages and reduce influence of misinformation or unfounded perception

areas planned for future development and in such locations (e.g. along the Thames and Humber estuaries) prices can be twice the average agricultural value.

Many factors limiting the implementation of managed realignment have been identified in this book and in the wider literature. Most limiting factors can be group into three categories: (1) governance and high-level strategy; (2) delivery of objectives at the project level (implementation and performance of projects); and (3) public perception and stakeholders engagement. To facilitate the implementation of managed realignment and increase the uptake by stakeholders and the public, the limitations within these three categories need to be addressed. Table 10.2 identifies generic requirements that will lead to improve implementation, understanding and acceptance of managed realignment. Not all requirements may be relevant to all methods of implementation or all individual projects.

References

Adams, C. A., Andrews, J. E., & Jickells, T. (2012). Nitrous oxide and methane fluxes vs. carbon, nitrogen and phosphorous burial in new intertidal and saltmarsh sediments. *Science of the Total Environment, 434,* 240–251.

Alexander, K. S., Ryan, A., & Measham, T. G. (2012). Managed retreat of coastal communities: Understanding responses to projected sea level rise. *Journal of Environmental Planning and Management, 55*(4), 409–433.

Almeida, D., Neto, C., Esteves, L. S., & Costa, J. C. (2014). The impacts of land-use changes on the recovery of saltmarshes in Portugal. *Ocean and Coastal Management, 92,* 40–49.

Burden, A., Garbutt, R. A., Evans, C. D., Jones, D. L., & Cooper, D. M. (2013). Carbon sequestration and biogeochemical cycling in a saltmarsh subject to coastal managed realignment. *Estuarine, Coastal and Shelf Science, 120,* 12–20.

Cearreta, A., García-Artola, A., Leorri, E., Irabien, M. J., & Masque, P. (2013). Recent environmental evolution of regenerated salt marshes in the southern Bay of Biscay: Anthropogenic evidences in their sedimentary record. *Journal of Marine Systems, 109*–110, 203–S212.

Committee on Climate Change. (2013). *Chapter 5: Regulating services—coastal habitats* (pp. 92–107). http://www.theccc.org.uk/wp-content/uploads/2013/07/ASC-2013-Chap5_singles_2.pdf.

Craft, C., Megonigal, P., Broome, S., Stevenson, J., Freese, R., Cornell, J., Zheng, L., Sacco, J. (2003). The pace of ecosystem development of constructed Spartina alterniflora marshes. *Ecological Applications, 13,* 1417–1432.

Davy, A. J., Brown, M. J. H., Mossman, H. L., & Grant, A. (2011). Colonization of a newly developing salt marsh: Disentangling independent effects of elevation and redox potential on halophytes. *Journal of Ecology, 99,* 1350–1357.

Esteves, L. S. (2013). Is managed realignment a sustainable long-term coastal management approach? *Journal of Coastal Research,* Special Issue *65*(1), 933–938.

Esteves, L. S., & Thomas, K. (2014). Managed realignment in practice in the UK: results from two independent surveys. *Journal of Coastal Research,* Special Issue 70, 407–413.

French, P. W. (2006). Managed realignment—The developing story of a comparatively new approach to soft engineering. *Estuarine, Coastal and Shelf Science, 67*(3), 409–423.

Garbutt, A., & Wolters, M. (2008). The natural regeneration of salt marsh on formerly reclaimed land. *Applied Vegetation Science, 11,* 335–344.

Goeldner-Gianella, L. (2007). Perceptions and attitudes toward de-polderisation in Europe: A comparison of five opinion surveys in France and the UK. *Journal of Coastal Research, 23*(5), 1218–1230.

Harman, B. P., Heyenga, S., Taylor, B. M., & Fletcher, C. S. (2014). Global lessons for adapting coastal communities to protect against storm surge inundation. *Journal of Coastal Research* (in press).

Ledoux, L., Cornell, S., O'Riordan, T., Harvey, R., & Banyard, L. (2005). Towards sustainable flood and coastal management: Identifying drivers of, and obstacles to, managed realignment. *Land Use Policy, 22*(2), 129–144.

Luisetti, T., Turner, R. K., Bateman, I. J., Morse-Jones, S., Adams, C. & Fonseca, L. (2011). Coastal and marine ecosystem services valuation for policy and management: Managed realignment case studies in England. *Ocean & Coastal Management, 54*(3), 212–224.

Marquiegui, M. A., & Aguirrezabalaga, F. (2009). Colonization process by macrobenthic infauna after a managed coastal realignment in the Bidasoa estuary (Bay of Biscay, NE Atlantic). *Estuarine, Coastal and Shelf Science, 84*, 598–604.

McGlashan, D. J. (2003). Managed relocation: An assessment of its feasibility as a coastal management option. *The Geographical Journal, 169*(1), 6–20.

Midgley, S., & McGlashan, D. J. (2004). Planning and management of a proposed managed realignment project: Bothkennar, Forth Estuary, Scotland. *Marine Policy, 28*, 429–435.

Morris, R. K. A. (2012). Managed realignment: A sediment management perspective. *Ocean & Coastal Management, 65*, 59–66.

Mossman, H. L., Davy, A. J., & Grant, A. (2012). Does managed coastal realignment create saltmarshes with 'equivalent biological characteristics' to natural reference sites? *Journal of Applied Ecology, 49*(6), 1446–1456.

Myatt-Bell, L. B., Scrimshaw, M. D., Lester, J. N., & Potts, J. S. (2002). Public perception of managed realignment: Brancaster West Marsh, North Norfolk, UK. *Marine Policy, 26*(1), 45–57.

Myatt, L. B., Scrimshaw, M. D., & Lester, J. N. (2003a). Public perceptions and attitudes towards a forthcoming managed realignment scheme: Freiston Shore, Lincolnshire, UK. *Ocean & Coastal Management, 46*(6–7), 565–582.

Myatt, L. B., Scrimshaw, M. D., & Lester, J. N. (2003b). Public perceptions and attitudes towards an established managed realignment scheme: Orplands, Essex, UK. *Journal of Environmental Management, 68*(2), 173–181.

Niven, R. J., & Bardsley, D. K. (2013). Planned retreat as a management response to coastal risk: A case study from the Fleurieu Peninsula, South Australia. *Regional Environmental Change, 13*, 193–209.

Parrott, A., & Burningham, H. (2008). Opportunities of, and constraints to, the use of intertidal agri-environment schemes for sustainable coastal defence: A case study of the Blackwater Estuary, southeast England. *Ocean & Coastal Management, 51*(4), 352–367.

Pethick, J. (2002). Estuarine and tidal wetland restoration in the United Kingdom: Policy versus practice. *Restoration Ecology, 10*(3), 431–437.

Reisinger, A., Lawrence, J., Hart, G., & Chapman, R. (2014). From coping to resilience: The role of managed retreat in highly developed coastal regions of New Zealand. In B. Glavovic, R. Kaye, M. Kelly, & A. Travers (Eds.), *Climate change and the coast: Building resilient communities*. CRC Press: London.

RICS (Royal Institution of Chartered Surveyors). (2013). *Rural Land Market Survey H1*. http://www.rics.org/uk/knowledge/market-analysis1/ricsrac-rural-market-survey/ricsrau-rural-land-market-survey-h1-2013/.

Roca, E., & Villares, M. (2012). Public perceptions of managed realignment strategies: The case study of the Ebro Delta in the Mediterranean basin. *Ocean & Coastal Management, 60*, 38–47.

Santín, C., de la Rosa, J. M., Knicker, H., Otero, X. L., Álvarez, M. Á, González-Vila, F. J. (2009). Effects of reclamation and regeneration processes on organic matter from estuarine soils and sediments. *Organic Geochemistry, 40*, 931–941.

SOGREAH. (2011). Etude de Faisabilite – Depolderisation partielle et eventuelle des Bas-Champs du Vimeu – La recherché d'un avenir sur un territoire perenne. Phase 1: Etat des lieux – diagnostic du territoire. Chapitre 8: Approche géographie sociale – perceptions.

Spencer, K. L., Cundy, A. B., Davies-Hearn, S., Hughes, R., Turner, S. & MacLeod, C. L. (2008). Physicochemical changes in sediments at Orplands Farm, Essex, UK following 8 years of managed realignment. *Estuarine, Coastal and Shelf Science*, *76*(3), 608–619.

Spencer, K. L., & Harvey, G. L. (2012). Understanding system disturbance and ecosystem services in restored saltmarshes: Integrating physical and biogeochemical processes. *Estuarine, Coastal and Shelf Science*, *106*, 23–32.

Taylor, B. M., & McAllister, R. R. J. (2013). Bringing it all together: Researcher dialogue to improve synthesis in regional climate adaptation in South-East Queensland, Australia. *Regional environmental change*. doi:10.1007/s10113-013-0517-4.

Wolters, M., Garbutt, A., & Bakker, J. P. (2005). Salt-marsh restoration: Evaluating the success of de-embankments in north-west Europe. *Biological Conservation*, *123*, 249–268.

Zedler, J. B., & Adam, P. (2002). Salt marshes. In M. R. Perrow & A. J. Davy (Eds.), *Handbook of ecological restoration. Restoration in practice* (Vol. 2, pp. 238–266). Cambridge: Cambridge University Press.

Chapter 11
Concluding Remarks

Luciana S. Esteves

Abstract This chapter summarises the main findings of previous chapters and identifies current experiences and future prospects concerning managed realignment. Increasingly managed realignment is seen as an alternative to traditional hard engineering with a capacity to deliver sustainable coastal management solutions that can account for climate change. The availability of suitable land and public acceptance are two important factors influencing the wider implementation of managed realignment. Recent national and regional strategies worldwide give managed realignment an increasing role in flood and erosion risk management; therefore, gaining stakeholders and public support is fundamental. The current understanding and uptake of managed realignment is hindered partly by the inconsistent use of the terminology. Adopting the definition proposed in Chap. 2 of this book, managed realignment is more widely implemented than first anticipated and experiences need to be more widely disseminated. Different regions and countries tend to adopt different managed realignment approaches. In the UK realignment of defences is a common managed realignment method; in Belgium controlled reduced tide is the preferred option; and in many countries the focus is on managed retreat (i.e. relocation of structures from high risk areas). Combining managed retreat and other managed realignment methods is the most climate-proof alternative to reduce people and property at risk from flooding and erosion. Fortunately, some countries are already working towards this long-term solution.

Managed realignment is a soft engineering approach for managing coastal erosion and flood risk. Increasingly it is seen as an alternative to traditional hard engineering with a capacity to deliver sustainable coastal management solutions that can account for climate change. In this respect, managed realignment aims to restore or maintain the adaptive capacity of natural environments in response to sea-level rise. Importantly, managed realignment has been widely implemented as a management approach to compensate for habitat loss due to coastal development and coastal squeeze.

L. S. Esteves (✉)
Faculty of Science and Technology, Bournemouth University, Talbot Campus,
Poole, Dorset, BH12 5BB, UK
e-mail: lesteves@bournemouth.ac.uk

L. S. Esteves, *Managed Realignment: A Viable Long-Term Coastal Management Strategy?*, 125
SpringerBriefs in Environmental Science, DOI 10.1007/978-94-017-9029-1_11,
© Springer Science+Business Media Dordrecht 2014

Underlying the implementation of managed realignment as a strategic climate change adaptation, is the potential to create opportunities for the development of fully functional coastal habitats able to benefit society through the provision of multiple ecosystem services. Most commonly, managed realignment project are designed to promote ecosystem services related to: natural protection against storms; flood control; and provision of habitat and biodiversity. Additionally, more recently attention has been given to the potential of managed realignment to provide carbon sequestration functions, nutrient cycling, water purification, recreation and amenity value.

The earliest managed realignment schemes were implemented in the 1980s and comprised of isolated initiatives to address local needs. A more strategic approach was taken from mid-1990s, when projects started to be implemented as part of estuary-wide plans. Since then a total of 125 managed realignment projects are currently known to have been implemented in Europe (see a list in the Appendix) and an unknown number of initiatives exist in the USA. Although practice has advanced knowledge in many technical aspects, it is surprising how little is still known about the long-term evolution of sites. This gap in knowledge is partially due to the fact that many projects are relatively recent, but also because systematic monitoring is inadequate or non-existent, data availability is restricted to organisations involved in the design and implementation of schemes, and only a few independent studies exist.

Managed realignment is still a novel approach to coastal management and examples to date are limited both in the number of projects and geographical spread. However, recent national and regional policies (e.g. in Belgium, France and the UK) have placed a strong emphasis on managed realignment approaches as a long-term strategy for coastal management. As a result, a large number of projects are planned to be implemented in the next decades. For example, by 2030 there are plans to realign about 10 % of the English and Welsh coastline length (see Sect. 4.1) and to create about 1300 ha of flood storage area and over 850 ha of new habitat along the Scheldt estuary at the border of Belgium and Netherlands (see Sect. 4.2). Planning for retreat from high risk areas is high in the agenda in France and New Zealand and increasingly in many other countries.

Despite the increased interest in managed realignment, there are a number of complicating factors which need to be addressed to facilitate its wider implementation in the future. Limiting factors identified in this book, include: inadequate political instruments (e.g. unclear targets at the strategic and individual project levels; inadequate funding mechanisms to purchase land or property; poor communication and public engagement strategies); technical viability (e.g. land availability, suitability of sites to create the desired habitats, environmental constraints; uncertainties due to variability of natural processes); poor knowledge of long-term evolution of sites (e.g. hydrodynamics and sedimentary processes, colonisation by biota, time-frame for mature ecosystems to develop; lack of or inadequate monitoring); and public perception.

Public opposition is the issue most often cited as causing delays or impeding implementation of managed realignment at the local level. Public opposition results from a combination of factors mainly related to: poor awareness about policy drivers and strategic targets; mistrust of government actions and actors; resistance to change from the status quo; lack of understanding about managed realignment; uncertainties about potential benefits; 'not in my backyard' attitude; and strong sectorial views. The three main elements contributing to negative public perception about managed realignment are: (1) confusing terminology and inconsistent definitions which make it difficult to understanding the general concept; (2) lack of clarity and consistency in the dissemination of strategic messages and objectives of individual projects; and (3) lack of evidence about benefits gained from existing projects.

Chapter 2 explains the inconsistencies and ambiguities in the use of terminology and proposes a wider definition of managed realignment that encompasses the many methods of implementation currently in use. Most often, managed realignment is defined as the planned breaching or removal of coastal defences to create new intertidal area in previously protected land, resulting in a landward realignment of the shoreline. Here it is suggested that managed realignment is used as a general term to reflect five methods of implementation: (1) removal of defences; (2) breach of defences; (3) realignment of defences; (4) controlled tidal restoration; and (5) managed retreat. It is considered that this revised definition will contribute to: (a) clarify the terminology to facilitate wider understanding of the central concept; (b) present managed realignment as a suite of approaches applicable to a range of environmental and socio-economic settings; and (c) demonstrate that, in its wider concept, managed realignment has been more widely implemented than currently anticipated (as many initiatives not currently recognised as managed realignment fit in the wider definition).

All these forms of managed realignment aim to improve flood and erosion risk management with added environmental value by creating space for the development of a more naturally functioning coast. However, in some cases (e.g. in the UK) there has been a strong emphasis on habitat creation and biodiversity and considerably less focus on the impact of managed realignment on flood and erosion risk to people and property. The mismatch between policy objectives and targets of individual projects (often unclear or omitting reduction of flood and erosion risk) has led to the public perception in the UK that managed realignment is a costly option to protect habitats and species at the expense of public safety and social interests. In response to this, there has been a recent call from practitioners and researchers for a more balanced coverage of the range of multiple functions and benefits associated with managed realignment.

Any change in policy affecting people (e.g. from hard engineering to managed realignment) must be supported by a consistent message, widely disseminated to local communities and stakeholders directly affected, which clarifies the rationale for the change. Engagement with stakeholders and local communities at early planning stages helps identifying the range of potential gains and losses from managed realignment projects at the local level. Project design can then be tailored to maximise the most desired benefits. Public engagement provides an opportunity to discuss expectations, including the time-frames in which benefits might be realised, and the effects of uncertainties related to long-term evolution of the sites.

It is of paramount importance to gain public support (and private funding) and to identify and quantify uncontested evidence of the benefits accrued by existing managed realignment schemes. Objective quantification can only be possible if the project performance is assessed against pre-defined targets and based on evidence obtained from consistent data collection and robust data analysis. Improvements in project design and performance can only be achieved through adaptive management, as implemented in the USA Coastal Wetlands Planning, Protection and Restoration Act (see Sect. 4.6). This approach involves: long-term (e.g. 20 years) systematic monitoring of existing projects; periodic review of performance based on evidence from monitoring; and project adjustments to improve performance where and when required. Lessons learned from each individual project are then applied to inform future projects.

There is little disagreement that conceptually, managed realignment has great potential to: (a) provide natural defence against storms and rising sea levels; (b) contribute to the achievement of wider environmental policy objectives (e.g. EU Habitats and Water Framework directives); and (c) deliver additional ecosystem services (e.g. recreation, carbon sequestration, fisheries etc.), which may vary according to site characteristics. It is now time to provide evidence that benefits are being realised through existing projects.

For managed realignment to realise its potential as a long-term sustainable coastal management alternative it is necessary to: (a) develop a long-term strategic plan that effectively integrates multiple objectives (e.g. habitat creation, flood protection and amenity); (b) clearly define local and national targets at specified time frames; (c) implement systematic monitoring so performance can be adequately measured against targets; and (d) review schemes periodically based on evidence so adjustments to each project and the overall strategy can be put in place where necessary.

In order to adapt to climate change, it is very important to devise a long-term strategy integrating the range of managed realignment methods and planning policies. Reducing the number of people and properties in areas at high risk of flooding and erosion is the only climate-proof option across all time-scales. Therefore, planning for managed retreat in combination with other managed realignment approaches might be the only sustainable option to reduce flood and erosion risk in low-lying coastal and estuarine areas. There is evidence in this book that this is an approach already on the agenda of concerned government and non-government organisations. Ultimately, as a society facing unknown climate change consequences, we cannot continue to live the way we have in the past. Government policies are already reflecting the need for radical change, and one of the great challenges now is to find a means of changing the attitude of individuals and communities towards climate change adaptation.

Appendix: List of Managed Realignment Projects in Europe

Existing managed realignment projects in Europe are listed below alphabetically per country and indicating the geographical location, method of implementation, year of project completion and project size. Note that few projects are listed for the USA, but the list is not intended to be comprehensive. The objective is to demonstrate that, although not widely recognised as managed realignment, some initiatives outside Europe fit the wider definition used in this book.

This list was produced based on evidence from existing literature and online sources, especially the Online Managed Realignment Guide[1] (OMReG). The OMReG provides a useful list of the managed realignment projects in Europe and their main characteristics. In February 2013, it listed a total of 102 projects in Europe, half of which in the UK. Information about the implementation of managed realignment projects is scattered and not always readily available. Some missing entries and inconsistencies found in the OMReG are corrected in the list provided here, which includes projects already implemented or under construction (as of February 2014). An online map of managed realignment projects in the UK is available from http://binged.it/1fFF9pF. This resource is intended to be reqularly updated and include projects in other countries in the future, providing up-to-date information about managed realignment projects.

For some of the projects listed in OMReG, no other source of information was found during the research conducted for the production of this book. Some projects, especially in the UK, Belgium, Netherlands and France, are documented in the grey literature and, a selected few, also in scientific publications. The main source of information for each project listed below is identified as a footnote (note that more sources might exist for some projects); when a footnote is missing, the OMReG is the main source of information.

Site name	Location	Method	Year	Area (ha)
Belgium				
Bergenmeersen[a]	Scheldt	Controlled tidal restoration—CRT	Works started in 2013	41
Dijlemonding[b]	Dijle estuary	Controlled tidal restoration—CRT	Works started in 2013	207

---

[1] The OMReG is an online catalogue (http://www.abpmer.net/omreg/) about managed realignment produced by ABPmer, a private environmental consultancy based in Southampton, UK.

L. S. Esteves, *Managed Realignment: A Viable Long-Term Coastal Management Strategy?*, 129
SpringerBriefs in Environmental Science, DOI 10.1007/978-94-017-9029-1,
© Springer Science+Business Media Dordrecht 2014

Site name	Location	Method	Year	Area (ha)
Hedwige-Prosper[59]	Scheldt	Realignment of defence	Works started in 2010	465
Heusden[c]	Scheldt	Breach of defence	2006	10
Kalkense Meersen[59]	Scheldt	Realignment of defence	Works started in 2013	606
Ketenisse schor[60]	Scheldt	Realignment of defence and land lowering	2003	60
Lippenbroek[59,d,e]	Scheldt	Controlled tidal restoration—CRT	2006	10
Paardebroek[60]	Scheldt	Realignment of defence	Works started in 2013	28
Paardenschor[60]	Scheldt	Realignment of defence and land lowering	2004	12
Paddebeek[60]	Scheldt	Realignment of defence	2003	1.6
Polders of Kruibeke[59]	Scheldt	Realignment of defence and controlled tidal restoration—CRT	Works started in 2013	650
Vlassenbroek and Wal-Zwijn[59]	Scheldt	Realignment of defence and controlled tidal restoration–CRT	Works started in 2012	416
Wijmeers 1[58]	Scheldt	Realignment of defence	Works started in 2013	159
Wijmeers 2[58]	Scheldt	Realignment of defence	Works started in 2013	28
Yzer mouth	Yzer	Removal of defence	2001	50
Denmark				
Geddal Strandenge	Limfjord	Breach of defence	1992	140
Viggelsø	Odense Fjord	Breach of defence	1993	66
France				
Aber de Crozon[f]	Crozon	Removal of defence	1981	87
Île Nouvelle[g]	Gironde estuary	Breach of defence	2000	265
Polder de Sébastopol[h]	Ile de Noirmouthier	Controlled tidal restoration	1999	132
Sète-Marseillan[i]	Languedoc-Roussillon	Managed retreat	2010	?
Germany				
Anklamer Stadtbruch	Oderhaff	Breach of defence	2004	1750
Beltringharder Koog[j]	North Friesland	Controlled tidal restoration	1988	853
Billwerder Insel	Elbe estuary	Removal of defence	2008	20
Dorumer Sommerpolder	East Friesland	Removal of defence	2001	4
Hahnöfer Sand	Elbe	Removal of defence	2002	104
Hauener Hooge	Ley Bay	Breach of defence	1994	80
Karrendorfer Wiesen	Greifswald Bodden	Removal of defence	1993	350
Kleinensieler Plate	Weser	Breach of defence	2000	58
Kleines Noor	Flensburg Fjord	Breach of defence	2002	14
Kreetsand[k]	Elbe	Realignment of defence (not yet breached)	1999 and 2012	30
Langeooger Sommerpolder[l]	Island of Langeoog	Removal of defence	2004	215

Site name	Location	Method	Year	Area (ha)
Luneplate	Weser	Controlled tidal restoration	2008	215
Lütetsburger Sommerpolder	Mainland coast of East Friesland	Removal of defence	1982	15
Pepelower/Tessmans-dorfer Wiesen	Salzhaff	Breach of defence	2002	120
Polder Freetz	Island of Rügen	Removal of defence	2002	180
Polder Friedrichsh-gen (Ziesetal)	Greifswald Bodden	Removal of defence	1999	90
Polder Neuensien (Südwestteil)	Island of Rügen	Breach of defence	2002	40
Polder Roggow	Salzhaff	Breach of defence	2002	40
Polder Wehrland	Peenestrom	Removal of defence	2004	113
Riepenburg	Elbe	Removal of defence	1995	1
Rönnebecker Sand	Weser	Breach of defence	2002	34
Salzwiesenprojekt Wurster Küste (Berensch/ Spieka-Neufeld)	Mainland coast of East Friesland	Controlled tidal restoration	1995	280
Sommerpolder Wurster Küste	Mainland coast of East Friesland	Breach of defence	2007	145
Spadenländer Spitze	Elbe	Removal of defence	2000	7.5
Strandseenlandschaft Schmoel	Kiel Bay	Breach of defence	1989	40
Tegeler Plate Polder	Weser	Breach of defence	1997	150
Teilpolder Waschow	Peenestrom	Removal of defence	2004	66
Vorder-/Hinterwerder Polder	Weser	Breach of defence	1997	30
Wrauster Bogen	Elbe	Removal of defence	1991	2.2
Netherlands				
Breebaart	Dollard	Controlled tidal restoration—CRT	2001	63
Bunkervallei, de Slufter	Island of Texel	Breach of defence	2002	3
De Kerf	North Holland	Breach of defence	1997	30
Groene Hoek, de Slufter	Island of Texel	Breach of defence	2002	13
Groene Strand	Island of Terschelling	Breach of defence	1996	23
Holwerder Zomerpolder	Mainland coast of Friesland	Breach of defence	1989	28
Klein Profijt	Oude Maas	Breach of defence	2005	6
Kroon's Polders	Island of Vlieland	Realignment of defence	1996	85
Noard Fryslân Bûtendyks	Friesland	Realignment of defence	2001	135
Tiengemeten	Haringvliet	Breach of defence	2007	500
Spain				
Marismas de la Vega de Jaitzubia	Bidasoa estuary	Removal of defence	2004	23

Site name	Location	Method	Year	Area (ha)
UK				
Abbotts Hall[m]	Blackwater	Controlled tidal restoration	1996	84
		Breach of defence	2002	84
Alkborough[n]	Humber	Breach of defence	2006	440
Alnmouth 1	Aln	Breach of defence (2 adjacent areas, one beach in each)	2006	8
Alnmouth 2	Aln	Breach of defence	2008	20
Amble Marshes[o]	Camel	Controlled tidal restoration	2010	56
Annery Kiln	Torridge	Breach of defence	2000	3.8
Barking Creek[p]	Barking Creek	Removal of defence	2006	0.3
Barking tidal barrier[c]	Barking Creek	Breach of defence	2006	1.0
Black Devon Wetlands[q]	Forth/Black Devon	Controlled tidal restoration	2000–2005	28
Black Hole Marsh	Axe	Controlled tidal restoration	2009	6
Bleadon Levels	Axe	Breach of defence	2001	13
Brancaster West Marsh[r]	North Norfolk	Realignment of defence	2002	7.5
Brandy Hole	Crouch	Breach of defence	2002	12
Chalkdock Marsh	Chichester Harbour	Controlled tidal restoration	2000	3.2
Chowder Ness[s]	Humber	Realignment of defence	2006	15
Clapper Marshes[b]	Camel	Controlled tidal restoration	2004–2011	10
Cone Pill	Severn	Breach of defence	2001	50
Devereux Farm[t]	Hamford Water	Breach of defence	2010	6
Freiston[u]	The Wash	Realignment of defence	2002	66
Glasson	Conder	Controlled tidal restoration	2005	6.4
Goswick Farm[b]	South Low River	Controlled tidal restoration	2010	4.5
Halvergate	Yare	Realignment of defence	2005	5
Havergate Island	Ore	Realignment of defence	2000	8.1
Hesketh Out Marsh[v]	Ribble Estuary	Realignment of defence	2008	168
Horsey Island	Hamford Water	Controlled tidal restoration	1995	1.2
Kennet Pans[w]	Firth of Forth	Realignment of defence	2008	8.2
Lepe/Darkwater[x]	Dark Water	Controlled tidal restoration	2007	5
London Gateway Wildlife Reserve	Thames	Realignment of defence	2010	27
Lower Clyst (Goosemoor)[y]	Exe estuary	Controlled tidal restoration	2004	6.2
Man Sands[z]	South Devon coast	Removal of defence	2004	2
Medmerry[aa]	West Sussex coast	Realignment of defence	2013	320
Millennium Terraces	Thames	Realignment of defence	1998	0.5

Site name	Location	Method	Year	Area (ha)
Nigg Bay[ab]	Cromarty Firth	Realignment of defence	2003	25
Northey Island	Blackwater	Realignment of defence	1991	0.8
Orplands[ac]	Blackwater	Realignment of defence	1995	38
Paull Holme Strays[e]	Humber	Realignment of defence	2003	80
Pillmouth[ad]	Torridge	Realignment of defence	2000	12.9
Ryan's Field	Hayle	Controlled tidal restoration	1995	6.23
Saltram (Blaxton Meadow)	Plym	Controlled tidal restoration	1995	5
Seal Sands (North-west Enclosure)	Tees	Controlled tidal restoration	1993	9
Skinflats Tidal Exchange Project—STEP Forth[ae]	Firth of Forth	Controlled tidal restoration	2009	11
Steart Peninsula[af]	Parrett estuary	Realignment of defence	2014 (expected)	416
Thorness Bay	The Solent	Breach of defence	2004	7
Thornham Point	Chichester Harbour	Breach of defence	1997	6.9
Tollesbury[ag]	Blackwater	Realignment of defence	1995	21
Treraven Meadows[b]	Camel	Controlled tidal restoration	2007	15
Trimley Marsh[ah]	Orwell	Breach of defence	2000	16.5
Upper Lantern Marsh[ai]	Ore	Breach of defence	1999	37
Vange Marsh	Thames	Controlled tidal restoration	2006	1
Walborough	Axe	Controlled tidal restoration	2004	4.5
Wallasea[aj]	Crouch	Realignment of defence	2006	115
Warkworth	Coquet	Breach of defence	2009	0.4
Watertown Farm	Yeo	Breach of defence	2000	1.5
Welwick[ak]	Humber	Realignment of defence	2006	54
USA				
Cheniere Ronquille Barrier Island Restoration[al]	Louisiana	Realignment of defence	2010	187
East Harbor Lagoon[am]	Cape Cod National Seashore	Controlled tidal restoration	2011	?
Isles Dernieres Restoration Trinity Island[an]	Louisiana	Realignment of defence	1999	314
Little River Marsh Restoration[ao]	North Hampton, New Hampshire	Controlled tidal restoration	2000	81
Whiskey Island Restoration[ap]	Louisiana	Realignment of defence	2004	420
Surfside Beach (9 houses relocated)	Brazoria County, Texas	Managed retreat	?	–
Tarramar Beach	City of Galveston, Texas	Managed retreat	?	–
Sunny Beach	West Galveston Island	Managed retreat	2010	–

Site name	Location	Method	Year	Area (ha)
Wailea Beach	Maui	Managed retreat	2012	–
Brigthon Beach Hotel	Coney Island, New York	Managed retreat	1888	–
Diamond City	North Carolina	Managed retreat	1899	–

[a] De Beukelaer-Dossche, M. and Decleyre, D. (eds.), 2013, www.vliz.be/imisdocs/publications/250657.pdf

[b] Sigma Plan, http://www.sigmaplan.be/en/

[c] Van den Bergh, E.; Verbessem, I.; Van den Neucker, T.; De Regge, N. and Soors, J., 2008. Evaluation of managed realignments in the Scheldt estuary. 6th European Conference on Ecological Restoration (Ghent, 8–12 Sept. 2008)

[d] Jacobs, S.; Beauchard, O.; Struyf, E.; Cox, T.; Maris, T. and Meire, P., 2009. Restoration of tidal freshwater vegetation using controlled reduced tide (CRT) along the Schelde Estuary (Belgium). *Estuarine, Coastal and Shelf Science*, 85, 368–376

[e] Beauchard, O., Jacobs, S., Cox, T. J. S., Maris, T., Vrebos, D., Van Braeckel, A., and Meire, P., 2011. A new technique for tidal habitat restoration: evaluation of its hydrological potentials. *Ecological Engineering*, 37, 1849–1858

[f] Digue Richet, Inventaire général du patrimoine culturel, http://patrimoine.region-bretagne.fr/sdx/sribzh/main.xsp?execute=show_document&id=MERIMEEIA29004203

[g] Conservatoire du Littoral, http://www.conservatoire-du-littoral.fr/siteLittoral/222/28-ile-nouvelle-33-_gironde.htm

[h] Pays de La Loire, RNR de Sébastopol, http://www.paysdelaloire.fr/les-pays-de-la-loire/reserves-regionales/actu-detaillee/n/polder-de-sebastopol/

[i] Heurtefeux, H.; Sauboua, P.; Lanzellotti, P. and Bichot, A., 2011. Coastal Risk Management Modes: The Managed Realignment as a Risk Conception More Integrated. In: Savino, M. (ed.), Risk Management in Environment, Production and Economy. InTech, Rijeka, Croacia

[j] RESTORE, 2013. Case study:Beltringharder Koog Regulated Tidal Exchange Scheme. Available at: http://restorerivers.eu/wiki/index.php?title=Case_study%3ABeltringharder_Koog_Regulated_Tidal_Exchange_Scheme

[k] IBA Hamburg, 2013. Pilot Project Kreetsand, http://www.iba-hamburg.de/en/themes-projects/elbe-islands-dyke-park/pilot-project-kreetsand/projekt/pilot-project-kreetsand.html

[l] Barkowski, J.W.; Kolditz, K.; Brumsack, H. and Freund, H., 2009. The impact of tidal inundation on salt marsh vegetation after de-embankment on Langeoog Island, Germany—six years time series of permanent plots. Journal of Coastal Conservation, 13,185–206

[m] Pickaver, A., 2010. A sustainable coastal defence re-creating wildlife habitats alongside economic farming methods, Abbott's Hall Farm—UK. OURCOAST project. Available from: http://ec.europa.eu/ourcoast/index.cfm?menuID=8&articleID=5, http://www.wwf.org.uk/filelibrary/pdf/mu52.pdf

[n] Manson, S. and Pinnington, N., 2012. Alkborough Managed Realignment (Humber estuary). Measure analysis in the framework of the Interreg IVB project TIDE. Measure 30. 25 pages. Hull. http://www.tide-toolbox.eu/pdf/measures/Alkborough.pdf

[o] Environment Agency, Camel Valley Wetland Restoration. http://www.environment-agency.gov.uk/static/documents/Leisure/Camel_Valley_Wetland_Restoration.pdf

[p] RESTORE Partnership (n/a), Lower River Roding Regeneration Project—Summary Report. http://restorerivers.eu/wiki/images/e/ef/Lower_River_Roding_Regeneration_Project.pdf

[q] Clackmannanshire Council, Black Devon Wetland, http://www.clacksweb.org.uk/environment/blackdevonwetland/

[r] Myatt, L.B.; Scrimshaw, M.D., Lester, J.N. and Potts, J.S., 2002. Public perceptions of managed realignment: Brancaster West marsh, North Norfolk, UK. Marine Policy, 26, 45–57

[s] Hemingway, K.L., Cutts, N.C. and R. Pérez-Dominguez., 2008. Managed Realignment in the Humber Estuary, UK. Institute of Estuarine & Coastal Studies (IECS), University of Hull, UK. Report produced as part of the European Interreg IIIB HARBASINS project. http://ec.europa.eu/ourcoast/download.cfm?fileID=863

[t] David Eagle (Deveraux Farm land owner), personal communication

[u] Friess, D.; Möller, I. and Spencer, T., 2008. Case study: managed realignment and the re-establishment of saltmarsh habitat, Freiston Shore, Linconlnshire, United Kingdom. ProAct Network, http://www.proactnetwork.org/proactwebsite/media/download/CCA_DRR_reports/casestudies/em.report.case_6.pdf

[v] Environment Agency, 2010. Working with natural processes to manage flood and coastal erosion risk. A guidance Document

[w] Scottish Executive Development Department, 2003. Kennet Pans Coastal Realignment Feasibility Study. http://www.transportscotland.gov.uk/files/documents/projects/upper-forth-crossing/UFC_Appendix6A.pdf

[x] Environment Agency, 2010. Working with natural processes to manage flood and coastal erosion risk. A guidance document, March 2010. http://cdn.environment-agency.gov.uk/geho0310bsfi-e-e.pdf

[y] Pickaver, A.H., 2010. Regulated tidal exchange as part of a broader strategy for managing marsh habitat, Goosemoor UK. OURCOAST project, http://ec.europa.eu/ourcoast/index.cfm?menuID=7&articleID=115

[z] Sout West Coastal Group, (undated). Managed Retreat. http://www.southwestcoastalgroup.org.uk/cc_defence_managedretreat.html

[aa] Environment Agency, Medmerry managed realignment scheme. http://www.environment-agency.gov.uk/homeandleisure/floods/109062.aspx

[ab] Crowther, A., 2007. The restoration of intertidal habitats for non-breeding waterbirds through breached managed realignment. PhD Thesis, University of Stirling, UK. https://dspace.stir.ac.uk/bitstream/1893/334/3/Thesis.pdf.txt

[ac] Spencer, K.L.; Cundy, A.B.; Davies-Hearn, S.; Hughes, R.; Turner, S. and MacLeod, C.L., 2008. Physicochemical changes in sediments at Orplands Farm, Essex, UK following 8 years of managed realignment. Estuarine, Coastal and Shelf Science, 76(3), 608–619

[ad] Blackwell, M.S.A.; Hogan, D.V. and Maltby, E. 2004. The short-term impact of managed realignment on soil environmental variables and hydrology. Estuarine, Coastal and Shelf Science, 59, 687–701

[ae] RESTORE, 2014. Case study: Welwick Managed Realignment Scheme. http://www.rspb.org.uk/media/releases/details.aspx?view=print&id=tcm:9-232808

[af] Environment Agency, 2011. Steart Coastal Management Project. Environmental Statement: non-technical summary. Report Reference code IMSW001389

[ag] Wolters, M.; Garbutt, A. and Bakker, J.P., 2005. Plant colonization after managed realignment: the relative importance of diaspore dispersal. Journal of Applied Ecology, 42(4), 770–777

[ah] John, S., 2007. Compensation habitat and port development: defining needs and demonstrating success. Haskoning UK, http://www.coastms.co.uk/resources/6d4af33e-6a68-4630-bb0c-643fa666f3cf.pdf

[ai] National Trust, Orford Ness National Nature Reserve, http://www.nationaltrust.org.uk/orford-ness/wildlife/view-page/item418077/

[aj] ABPmer, 2013. Wallasea Wetlands Creation Project. http://www.abpmer.net/wallasea/

[ak] http://restorerivers.eu/wiki/index.php?title=Case_study%3AWelwick_Managed_Realignment_Scheme

[al] CWPPRA projects, http://lacoast.gov/reports/gpfs/BA-76.pdf

[am] Smith, S. and Medeiros, K., 2013. Manipulation of water levels to facilitate vegetation change in a coastal lagoon undergoing partial tidal restoration (Cape Cod, Massachusetts). Journal of Coastal Research, 29(6A), 93–99

[an] CWPPRA projects, http://lacoast.gov/new/Projects/Info.aspx?num=TE-24

[ao] NOAA Restoration Center and NOAA Coastal Services Center, 2010. Returning the Tide: A Tidal Hydrology Restoration Guidance Manual for the Southeastern United States. NOAA, Silver Spring. Available at: http://www.habitat.noaa.gov/partners/toolkits/

[ap] CWPPRA projects, http://lacoast.gov/new/Projects/Info.aspx?num=TE-50

Index

L. S. Esteves, *Managed Realignment: A Viable Long-Term Coastal Management Strategy?*,   137
SpringerBriefs in Environmental Science, DOI 10.1007/978-94-017-9029-1,
© Springer Science+Business Media Dordrecht 2014

Printed by Printforce, the Netherlands